Coimbra Mathematical Texts

Volume 4

The Mathematics Department of the University of Coimbra, one of the oldest universities in the world, has a new collection of advanced level mathematical texts, *Coimbra Mathematical Texts*, in collaboration with Springer Nature. This collection aims to publish monographs and conference proceedings of excellent quality, in all branches of mathematics, that will inspire the future generations.

Any researcher interested to contribute to this collection should send an e-mail to cmt@mat.uc.pt with subject “Possible contribution”.

The editorial team will perform an initial appraisal of the submitted manuscript based on the quality of the submission, interest and importance of the topic. If the contribution is considered suitable to be sent to peer review, it will be reviewed by specialists of international reputation.

Miguel Abreu · Henrique Bursztyn · Jorge Picado
Editors

Brazil-Portugal Mathematics

Proceedings of the First Brazil-Portugal Joint Meeting in Mathematics, Salvador, Brazil, August 14–20, 2022

Springer

Editors
Miguel Abreu
Departamento de Matemática
Instituto Superior Técnico
Lisbon, Portugal

Henrique Bursztyn
Instituto Nacional de Matemática
Pura e Aplicada
Rio de Janeiro, Brazil

Jorge Picado
Departamento de Matemática
Universidade de Coimbra
Coimbra, Portugal

ISSN 2813-0057 ISSN 2813-0065 (electronic)
Coimbra Mathematical Texts
ISBN 978-3-031-98454-9 ISBN 978-3-031-98455-6 (eBook)
https://doi.org/10.1007/978-3-031-98455-6

This Springer imprint is published by the registered company Springer Nature Switzerland AG
The registered company address is: Gewerbestrasse 11, 6330 Cham, Switzerland

Preface

The content of this book is based on the plenary lectures delivered at the First Brazil-Portugal joint meeting in Mathematics, held at Universidade Federal da Bahia, from August 14 to 20, 2022. Hosting this event in Salvador, Brazil's first capital from 1549 to 1763, marked the bicentenary of Brazil's independence and celebrated the growing scientific interactions between Brazil and Portugal. The meeting was conceived as a starting point for a new phase of bilateral cooperation in mathematics.

Promoted by the Brazilian Mathematical Society (SBM), the Portuguese Mathematical Society (SPM) and the Brazilian Society of Applied and Computational Mathematics (SBMAC), the event brought together 200 registered participants that took part in 19 thematic sessions representing different areas of collaboration between the Portuguese and Brazilian mathematical communities. The conference was organized by Elen Barbosa (UFBA), Evandro Santos (UFBA), Henrique Bursztyn (IMPA), Joilson Ribeiro (UFBA), Kleyber Cunha (UFBA), Paulo Varandas (UFBA and FCT-CMUP), Suely Lima (IMPA), and Vilton Pinheiro (UFBA).

The main subjects covered in this book include:

- classification of real and complex singularities;
- transmission problems in mathematical analysis;
- complexity in continuous optimization;
- statistical extreme value theory;
- periodic orbits of autonomous Hamiltonian systems;
- fractional PDEs from interacting particle systems;
- deformation quantization of Poisson manifolds.

We would like to express our gratitude to the authors (and their collaborators) for contributing extended versions of their lecture notes to this volume:

- Ernesto Birgin (Universidade de São Paulo),
- Alexandre Fernandes (Universidade Federal do Ceará),
- Rui Loja Fernandes (University of Illinois at Urbana-Champaign) and Alejandro Cabrera (Universidade Federal do Rio de Janeiro),
- Maria Ivette Gomes (FC—Universidade de Lisboa),

– Patrícia Gonçalves (IST—Universidade de Lisboa) and Pedro Cardoso (University of Bonn),
– Leonardo Macarini (IST—Universidade de Lisboa),
– Edgard Pimentel (Universidade de Coimbra), Vincenzo Bianca (Universidade de Coimbra) and José Miguel Urbano (Universidade de Coimbra and KAUST).

Their invaluable contributions, alongside the dedicated work of the referees, have made this publication possible.

We are equally grateful to the editors of the *Coimbra Mathematical Texts* series—Ana Paula Santana, Júlio S. Neves, Marcelo Viana, Maria Paula Serra de Oliveira, and Rui Loja Fernandes—for their continuous editorial support throughout this process. We also extend our thanks to the members of the Scientific Committee of the conference for their guidance and contributions to the conference:

– Henrique Bursztyn (IMPA, Rio de Janeiro),
– José Alberto Cuminato (Universidade de São Paulo),
– José Ferreira Alves (Universidade do Porto),
– José Francisco Rodrigues (FC—Universidade de Lisboa),
– Marcio Gomes Soares (Universidade Federal de Minas Gerais),
– Maria Manuel Clementino (Universidade de Coimbra),
– Miguel Abreu (IST—Universidade de Lisboa), and
– Nancy Garcia (Universidade Estadual de Campinas).

Finally, we gratefully acknowledge the generous support of the following institutions that made possible the organization of the meeting leading to the publication of this book:

– Universidade Federal da Bahia;
– Instituto de Matemática Pura e Aplicada, Rio de Janeiro;
– Sociedade Brasileira de Matemática;
– Sociedade Portuguesa de Matemática;
– Sociedade Brasileira de Matemática Aplicada e Computacional;
– CAPES, Brazil;
– Fundação para a Ciência e a Tecnologia, Portugal.

Lisbon, Portugal
Rio de Janeiro, Brazil
Coimbra, Portugal
September 2024

Miguel Abreu
Henrique Bursztyn
Jorge Picado

Contents

Transmission Problems: Regularity Theory, Interfaces and Beyond

Vincenzo Bianca, Edgard A. Pimentel, and José Miguel Urbano

Abstract Modelling diffusion processes in heterogeneous media requires addressing inherent discontinuities across interfaces where specific conditions must be met. These challenges fall under the purview of Mathematical Analysis as *transmission problems*. We present a panorama of the theory of transmission problems, encompassing the seminal contributions from the 1950s and subsequent developments. Then, we delve into the discussion of regularity issues, including recent advances matching the minimal regularity requirements of interfaces and the optimal regularity of the solutions. A discussion on free transmission problems closes the survey.

1 Introduction

The study of the physical properties of heterogeneous media plays a fundamental role across disciplines, both in the basic sciences and in their applications. An underlying phenomenon governing an important set of those properties is *diffusion*. Given the heterogeneity of the media, it is reasonable to consider the diffusion process changes across the different components of the medium. Since the second half of the last century, methods in the realm of partial differential equations (PDE, for short) and the calculus of variations have been used to examine these changes. Their various mathematical formulations are known in the literature as *transmission problems*.

The general setting is the following. Let $\Omega \subseteq \mathbb{R}^d$ be a bounded open set; for $k \in \mathbb{N}$ fixed, consider $\Omega_1, \ldots, \Omega_k \Subset \Omega$, pairwise disjoint. Suppose $u : \Omega \to \mathbb{R}$ solves within every Ω_i, $i = 1, \ldots, k$, some prescribed PDE. A natural question concerns

V. Bianca · E. A. Pimentel (✉) · J. M. Urbano
University of Coimbra, CMUC, Department of Mathematics, Coimbra, Portugal
e-mail: edgard.pimentel@mat.uc.pt

V. Bianca
e-mail: vincenzo@mat.uc.pt

J. M. Urbano
King Abdullah University of Science and Technology (KAUST), Computer, Electrical and Mathematical Sciences and Engineering Division (CEMSE), Thuwal, Saudi Arabia
e-mail: miguel.urbano@kaust.edu.sa

M. Abreu et al. (eds.), *Brazil-Portugal Mathematics*, Coimbra Mathematical Texts 4,
https://doi.org/10.1007/978-3-031-98455-6_1

the interactions of the k restrictions

$$u_i := u\big|_{\Omega_i}, \qquad \text{for} \qquad i = 1, \ldots, k,$$

with each other.

At this point, an important distinction is in place. In fact, one must distinguish the case of an interface that is fixed a priori from the case of a solution-dependent interface. The latter is referred to as *free interface* or *free boundary*. From a chronological viewpoint, fixed transmission problems precede the free interface variant.

Transmission problems first appeared in Mathematical Analysis in the work of Mauro Picone, circa 1950 [49]. The author addresses a fixed transmission problem arising in elasticity theory in his seminal contribution. The main result in [49] is the uniqueness of solutions for the transmission problem. The author also indicates a strategy aiming at establishing existence, which, however, is not explored at length in the paper.

The work of Mauro Picone is consequential, as it attracted immediate attention from the mathematical community to this class of problems. We highlight the work of Jacques-Louis Lions [45], where the first rigorous theory of the existence of solutions is put forward. Among the subsequent studies inspired by Picone's ideas, we mention [14, 21–23, 36, 46, 48, 55]. We refer the reader to the neatly written monograph [13], where a detailed account of the theory of fixed transmission problems and a comprehensive list of references are available.

Once the questions on the existence and uniqueness of solutions have been well-understood and documented, the community's attention shifted to the regularity theory. A pioneering contribution in this regard is in the work of Yanyan Li and Michael Vogelius [43]. In that paper, the authors consider a domain $\Omega \subseteq \mathbb{R}^d$, split into $k \in \mathbb{N}$ pairwise disjoint subdomains $\Omega_1, \ldots, \Omega_k \Subset \Omega$. The mathematical model is driven by the PDE

$$\frac{\partial}{\partial x_i}\left(a(x)\frac{\partial u}{\partial x_j}\right) = f \quad \text{in} \quad \Omega,$$

where $f : \Omega \to \mathbb{R}$ satisfies usual continuity and integrability assumptions, and

$$a(x) = \begin{cases} a_i(x) & \text{for} \ \ \Omega_i, \ i = 1, \ldots, k \\ a_{k+1}(x) & \text{for} \ \ \Omega \setminus \cup_{i=1}^k \Omega_i. \end{cases}$$

The main contribution of [43] is the local Hölder-continuity for the gradient of the solutions. The vectorial counterpart of the results in [43] appeared in the work of Yanyan Li and Louis Nirenberg [42], where the authors also establish estimates for higher-order derivatives of the solutions. From the perspective of applications, this model accounts for different inclusions within a homogeneous medium, the primary example being fibre-reinforced materials.

The regularity analysis in the context of conductivity and insulation is the subject of [3, 4]. Here, the authors consider a bounded domain $\Omega \subseteq \mathbb{R}^d$, and two disjoint subdomains $\Omega_1, \Omega_2 \Subset \Omega$ that are ε-apart, that is $\operatorname{dist}(\Omega_1, \Omega_2) = \varepsilon$. Within each subdomain, we are in the presence of an equation in divergence form. The analysis of the problem leads to estimates which are dependent on the parameter $\varepsilon > 0$. Indeed, they ensure a deterioration on the bounds for the gradient, which blows up as $\varepsilon \to 0^+$.

In the former literature, the program has focused on the existence and regularity of the solutions under regular enough fixed interfaces. Recently, the fundamental question of examining the *minimal* geometric requirements on the fixed interface to recover regularity estimates on the solutions of the fixed transmission problem has been addressed. This is the analysis in the work of Luis Caffarelli, María Soria-Carro and Pablo Stinga [20], where the authors consider a bounded domain $\Omega \subseteq \mathbb{R}^d$, and fix $\Omega_1 \Subset \Omega$, defining $\Omega_2 := \Omega \setminus \overline{\Omega_1}$. The key assumption in [20] concerns the geometry of $\Gamma := \partial\Omega_1$, which the authors assume to be of class $C^{1,\alpha}$.

The problem is to look for a function $u \in W^{1,2}_{\text{loc}}(\Omega)$ whose restrictions $u_i := u\big|_{\overline{\Omega_i}}$ satisfy

$$\begin{cases} \Delta u_i = 0 & \text{in } \Omega_i, \quad \text{for } i = 1, 2, \\ u_2 = 0 & \text{in } \partial\Omega, \\ (u_1)_\nu - (u_2)_\nu = g & \text{on } \Gamma. \end{cases}$$

By resorting to well-known properties of harmonic functions, the authors prove the existence and uniqueness of the solutions, together with a regularity result in $C^{0,\text{Log-Lip}}$-spaces *across* the fixed interface Γ.

However, the main contribution is [20] concerns the regularity of solutions *up to the interface*. The authors develop a new stability result relating general $C^{1,\alpha}$-regular interfaces with flat ones. Their approach relies on a clever combination of the mean value formula and the maximum principle for harmonic functions. As a result, they show that $u_i \in C^{1,\alpha}_{\text{loc}}(\overline{\Omega_i})$ for $i = 1, 2$.

In [54], the authors extend the analysis put forward in [20] to the context of fully nonlinear equations. In that paper, a complete theory of viscosity solutions to fully nonlinear fixed transmission problems is developed, including existence and uniqueness, comparison principles, and regularity estimates. As in [20], the authors establish the regularity of the solutions up to the fixed interface; we notice the regularity regime for the solutions matches the geometric properties of the fixed interface. Regarding the $C^{1,\alpha}$-regularity of solutions up to the interface, we refer the reader to [29] for a variant of the proof put forward in [20]. The argument in [29] does not rely either on the mean value property or on the maximum principle for harmonic functions. The paper also includes extensions to the context of $C^{1,\text{Dini}}$-fixed interfaces.

This introduction has covered so far some of the developments in the context of *fixed transmission problem*. That is, in the presence of interfaces that are determined *a priori*. An important variant of the topic regards the so-called *free transmission problems*, where the interface depends on the solution of the problem. The first attempt to examine a free transmission problem appeared in [1], where the authors propose a variational model. More precisely, they consider maps $A_\pm : \Omega \to S(d)$, where $S(d)$ stands for the space of symmetric matrices of order d, and study the functional

$$I(v) := \int_\Omega \frac{1}{2} \langle A(v,x)Dv, Dv\rangle + \Lambda(v,x) + fv \, \mathrm{d}x, \tag{1}$$

where the matrix $A(v,x)$ and the function $\Lambda(v,x)$ are defined as

$$A(v,x) := A_+(x)\chi_{\{v>0\}} + A_-(x)\chi_{\{v<0\}}$$

and

$$\Lambda(v,x) := \lambda_+(x)\chi_{\{v>0\}} + \lambda_-(x)\chi_{\{v<0\}},$$

and $f \in L^\infty(\Omega)$. The program pursued in [1] includes the existence of minimizers for (1), together with regularity results in the spaces of Hölder and Lipschitz-continuous functions.

In the context of free transmission problems at the intersection of PDE and the calculus of variations, we mention the recent work of Maria Colombo, Sungham Kim and Henrik Shahgholian [25]. The authors consider the functional

$$J(v,\Omega) := \int_\Omega \left(|Dv^+|^p + |Dv^-|^q\right) \mathrm{d}x,$$

where $1 < p, q < \infty$, with $p \neq q$. Their findings include properties of minimizers as well as information on the associated free boundary. They prove the existence of minimizers and their local Hölder continuity, with estimates. Concerning the analysis of the free boundary, the authors derive a free boundary condition and prove the free boundary is of class $C^{1,\alpha}$, almost everywhere, with respect to the measure $\Delta_p u^+$. The paper also verifies the support of $\Delta_p u^+$ is of σ-finite $(n-1)$-dimensional Hausdorff measure.

In the realm of fully nonlinear operators, free transmission problems have been studied from two different perspectives, namely the existence of generalized solutions (in the viscosity and the strong senses) and their regularity properties (see [35, 51]). In the uniformly elliptic problem, the model under consideration has the form

$$F_1(D^2u)\chi_{\{u>0\}} + F_2(D^2u)\chi_{\{u<0\}} = f \quad \text{in} \quad \Omega \cap (\{u>0\} \cup \{u<0\}), \tag{2}$$

where $F_i : S(d) \to \mathbb{R}$ are fully nonlinear uniformly elliptic operators and $f \in L^\infty(\Omega)$ is a given source term.

The existence of solutions for the Dirichlet problem associated with (2), both in the L^p-viscosity and in the L^p-strong sense, was established in [51]. The regularity of the solutions is the subject of [50].

A degenerate variant of (2) has also been examined in the literature. Let $0 < p_1 < p_2$ be fixed positive numbers and consider

$$|Du|^{p_1\chi_{\{u>0\}} + p_2\chi_{\{u<0\}}} F(D^2u) = f \quad \text{in} \quad \Omega \cap (\{u>0\} \cup \{u<0\}). \tag{3}$$

The model in (3) is inspired by the tradition of fully nonlinear equations degenerating as a power of the gradient; see [2, 7–12, 37], among others. In [35], the authors prove the existence of solutions to (3) and examine their regularity properties. In particular, they prove a $C^{1,\alpha}$-regularity estimate. Finally, the modulus of continuity depends explicitly on the Krylov-Safonov theory available for $F = 0$ and on the degeneracy rates $0 < p_1 < p_2$.

The variant of (3) with nonhomogeneous degeneracies is the subject of [27]. In that paper, the author considers a model of the form

$$H(x, u, Du)F(D^2u) = f \quad \text{in} \quad \Omega,$$

where F is uniformly elliptic, $f \in C(\Omega) \cap L^\infty(\Omega)$, and

$$H(x, u, Du) := |Du|^{p_1 \chi_{\{u>0\}} + p_2 \chi_{\{u<0\}}} + a(x)\chi_{\{u>0\}}|Du|^q + b(x)\chi_{\{u<0\}}|Du|^s,$$

for nonnegative functions $a, b \in C(\Omega)$, and nonnegative exponents p_1, p_2, q, and s. The findings in [27] cover the existence of solutions and $C^{1,\alpha}$-regularity estimates.

The remainder of this chapter is organized as follows. Section 2 discusses a few examples of fixed transmission problems, covering some of the main aspects of the theory. Section 2.1 describes the seminal work of Mauro Picone [49] and presents an argument due to Jacques-Louis Lions for the existence of weak solutions to Picone's problem. More recent results on regularity theory are the subject of Sect. 2.2. In Sect. 2.3, we examine the regularity theory of fixed transmission problems in the presence of minimal geometric requirements on the transmission interface. We describe quasilinear degenerate problems in Sect. 3, whereas their regularity in borderline spaces is the subject of Sect. 3.1. The chapter ends with a discussion on free transmission problems in Sect. 4.

2 An Overview on Fixed Transmission Problems

2.1 Laying the Foundations: First Models and a Theory of Existence

In this section, we describe the foundations of the theory of transmission problems in detail. In $\mathbb{R}^d$, with $d \geq 3$, let Ω_1 and Ω_2 be two disjoint open sets. Suppose that Ω_2 is the complement of a compact set. Suppose that $\partial\Omega_1$ and $\partial\Omega_2$ share an $(d-1)$-dimensional manifold Γ, continuously differentiable except, at most, at a finite set of points. For convenience, set $S_i := \partial\Omega_i \setminus \Gamma$, for $i = 1, 2$, and

$$\Omega := \Omega_1 \cup \Omega_2. \tag{4}$$

Now, let h_1, k_1 (resp. h_2, k_2) be the constants of Lamé of Ω_1 (resp. Ω_2). Lamé constants are two material-dependent quantities that arise in strain-stress relationships.

These parameters are named after Gabriel Lamé.[1] One looks for two vector fields, U^1 and U^2, defined in Ω_1, and Ω_2 respectively, with values in $\mathbb{R}^d$, such that

$$\begin{cases} h_1 \Delta U^1 + (h_1 + k_1) D(\operatorname{div} U^1) + f^1 = 0 & \text{in } \Omega_1 \\ h_2 \Delta U^2 + (h_2 + k_2) D(\operatorname{div} U^2) + f^2 = 0 & \text{in } \Omega_2, \end{cases} \tag{5}$$

under the interface conditions

$$\begin{cases} U^1 = U^2 & \text{on } \Gamma \\ t_1(U^1) + t_2(U^2) = 0 & \text{on } \Gamma, \end{cases} \tag{6}$$

where $D(\operatorname{div} U^i)$ is the gradient of the divergence of U^i for $i = 1, 2$.

Here, f^1 and f^2 are two given vector fields, representing external forces, and $t_1(U^1)$, $t_2(U^2)$ are the pressure fields on $\partial\Omega_1$ and $\partial\Omega_2$, respectively. Moreover, $t_i(U^i)$ has to be equal to the vector field on S_i, for $i = 1, 2$, and $U^2(x) \to 0$ as $|x| \to \infty$. In [49], Picone proved the uniqueness of the solution to this problem, claiming that solutions exist. Our choice in this survey is to present the general lines of an existence theory for (5)–(6), due to Jacques-Louis Lions [45].

We need some ingredients from the theory of functional spaces to discuss the argument put forward by Jacques-Louis Lions leading to the existence of a solution to (5)–(6). Let $\Omega \subseteq \mathbb{R}^d$ be bounded and open. For $u \in W^{1,2}(\Omega; \mathbb{R}^d)$, with $k, \ell \in \{1, \ldots, d\}$, define

$$s_{k\ell}(u) := \frac{1}{2}\left(\frac{\partial u_k}{\partial x_\ell} - \frac{\partial u_\ell}{\partial x_k}\right)$$

and

$$S(u) := \sum_{k,\ell=1}^{d} \|s_{k\ell}(u)\|^2_{L^2(\Omega)}.$$

For $u, v \in L^2(\Omega; \mathbb{R}^d)$, we denote by $(u, v)_{L^2(\Omega)}$ the inner product between u and v in $L^2(\Omega; \mathbb{R}^d)$, i.e.,

$$(u, v)_{L^2(\Omega)} := \int_\Omega u \cdot v \, \mathrm{d}x$$

For $u, v \in W^{1,2}(\Omega; \mathbb{R}^d)$, we then set

$$S(u, v) := \sum_{k,\ell=1}^{d} \big(s_{k\ell}(u), s_{k\ell}(v)\big)_{L^2(\Omega)}.$$

[1] Gabriel Lamé was a French mathematician born in Tours in 1795. With various contributions to a wide range of topics in Mathematics, he is known for a theory of curvilinear coordinates and the study of ellipse-like curves, currently referred to as *Lamé curves*. A Foreign Member of the Royal Swedish Academy of Sciences, elected in 1854, Lamé is one of the 72 names inscribed in the Eiffel Tower. He died in Paris in 1870.

Note that $S(u) = S(u, u)$, and also

$$S(u) \leq \sum_{k,\ell=1}^{d} \left\| \frac{\partial u_k}{\partial x_\ell} \right\|_{L^2(\Omega)}^2. \tag{7}$$

The definition of the following space will play a central role in the sequel. We recall the Sobolev exponent $2^* > 2$ is given by

$$2^* := \frac{2d}{d-2},$$

whereas its Hölder conjugate is

$$(2^*)' := \frac{2d}{d+2}.$$

Definition 1 Let Ω be as in (4). For $i = 1, 2$, let $T : W^{1,2}(\Omega_i; \mathbb{R}^d) \to L^2(\Gamma; \mathbb{R}^d)$ be the trace operator. We define V to be the space of functions belonging to $L^{2^*}(\Omega; \mathbb{R}^d) \cap W^{1,2}(\Omega; \mathbb{R}^d)$ such that $T(u^1) = T(u^2)$, where $u^1 \in W^{1,2}(\Omega_1; \mathbb{R}^d)$ and $u^2 \in W^{1,2}(\Omega_2; \mathbb{R}^d)$.

One fundamental step in the reasoning put forward in [45] is to prove the inequality in (7) can be reversed, up to constant factors, in V. Thanks to this, V will be a Hilbert space endowed with a natural norm, and the existence of an isomorphism between a suitable subspace of V and $L^{\frac{2d}{d+2}}(\Omega; \mathbb{R}^d)$ will lead to the existence of a unique solution. Before stating a result on the existence of solutions to (5)–(6), we detail two definitions concerning open and connected subsets of $\mathbb{R}^d$.

Definition 2 (*Sobolev set*) Let $\Omega \subseteq \mathbb{R}^d$ be connected and open. If $|\Omega| < \infty$, we say that Ω is a Sobolev set if $W^{1,2}(\Omega) \subset L^{2^*}(\Omega)$. If $|\Omega| = \infty$, we say that Ω is a Sobolev set if for every $u \in W^{1,2}(\Omega)$ there exists a constant $c = c(u)$ such that $u + c(u) \in L^{2^*}(\Omega)$.

Remark 3 (*Sobolev sets and the geometry of domains*) We notice that if the boundary of Ω is locally Lipschitz-regular, the inclusion $W^{1,2}(\Omega) \subset L^{2^*}(\Omega)$ follows from standard results in the theory of Sobolev spaces. As a consequence, every Lipschitz-regular domain is a Sobolev set. However, the notion of *Sobolev set* also accommodates less regular subsets of $\mathbb{R}^d$ for which that inclusion is not available.

Definition 4 (*Friedrichs set*) Let $\Omega \subseteq \mathbb{R}^d$ be connected and open. We say that Ω is a Friedrichs set if there exist $K \subseteq \Omega$, compact, and a constant $C \in (0, 1)$ such that, for every $u \in W^{1,2}(\Omega; \mathbb{R}^d)$ satisfying

$$-\Delta u - D(\operatorname{div} u) = 0 \quad \text{in} \quad \Omega,$$

we have

$$S(u) \leq \frac{1}{1-C} S_K(u),$$

where

$$S_K(u) := \sum_{k,\ell=1}^{d} \|s_{k\ell}\|^2_{L^2(K)}.$$

We refer the reader to [32, 33] for further context on Friedrichs sets. The main result in [45] is the following theorem.

Theorem 5 (Existence of a unique solution) *Let $\Omega \subseteq \mathbb{R}^d$ be as in* (4). *Suppose that Ω_1 and Ω_2 are Sobolev and Friedrichs sets. Then there exists a unique solution to the problem of Picone* (5)–(6).

Now, we continue by detailing the general lines of the proof of Theorem 5. We proceed with a lemma related to the space of tempered distributions. First, equip the linear space $L^{2^*}(\Omega; \mathbb{R}^d) \cap W^{1,2}(\Omega; \mathbb{R}^d)$ with the norm

$$\|\cdot\|_{2^*,2,\Omega} : L^{2^*}(\Omega; \mathbb{R}^d) \cap W^{1,2}(\Omega; \mathbb{R}^d) \to [0, \infty),$$

given by

$$\|u\|_{2^*,2,\Omega} := \|u\|_{L^{2^*}(\Omega;\, \mathbb{R}^d)} + \|Du\|_{L^2(\Omega;\, \mathbb{R}^d)}\,.$$

We denote with $\mathcal{D}_{2^*,2}(\Omega; \mathbb{R}^d)$ the closure of $\mathcal{D}(\Omega; \mathbb{R}^d)$ in the space $L^{2^*}(\Omega; \mathbb{R}^d) \cap W^{1,2}(\Omega; \mathbb{R}^d)$, with respect to the norm $\|\cdot\|_{2^*,2,\Omega}$.

Lemma 6 *For every $u \in \mathcal{D}_{2^*,2}(\Omega; \mathbb{R}^d)$ we have*

$$\sum_{k,\ell=1}^{d} \left\| \frac{\partial u_k}{\partial x_\ell} \right\|^2_{L^2(\Omega)} \leq 2S(u).$$

The previous result, whose proof follows by a simple integration by parts and by resorting to a density argument, says that the inequality in (7) can be reversed in $\mathcal{D}_{2^*,2}(\Omega; \mathbb{R}^d)$. This is a preliminary step in proving one can reverse (7) in the space V.

Define the bilinear form $\langle \cdot, \cdot \rangle : W^{1,2}(\Omega; \mathbb{R}^d) \to \mathbb{R}$ as

$$\langle u, v \rangle := 2hS(u, v) + k(\operatorname{div} u, \operatorname{div} v)_{L^2(\Omega)}$$

with $h > 0$ and $3k + 2h > 0$. The operator L defined as

$$L := -h\Delta - (h + k)D\mathrm{div} \tag{8}$$

is such that

$$\langle u, v \rangle = \langle Lu, v \rangle_{(\mathcal{D}_{2^*,2}(\Omega;\mathbb{R}^d)', \mathcal{D}_{2^*,2}(\Omega;\mathbb{R}^d))}\,,$$

where the crochet at the right-hand side represents the duality between $\mathcal{D}_{2^*,2}(\Omega;\mathbb{R}^d)'$ and $\mathcal{D}_{2^*,2}(\Omega;\mathbb{R}^d)$. The operator (8) is elliptic, i.e., there exists $a>0$ such that

$$\langle u,u\rangle \geq aS(u),$$

for every $u\in\mathcal{D}(\Omega;\mathbb{R}^d)$.

Because the operator in (8) is elliptic, it defines an isomorphism between $\mathcal{D}_{2^*,2}(\Omega;\mathbb{R}^d)'$ and $\mathcal{D}_{2^*,2}(\Omega;\mathbb{R}^d)$. An important ingredient in the proof of Theorem 5 concerns extending Lemma 6 to functions in the functional space V. This is the content of the next proposition.

Proposition 7 *Let $\Omega\subseteq\mathbb{R}^d$ be as in (4). Suppose that Ω_1 and Ω_2 are both Sobolev and Friedrichs sets. Then, there exists $C>0$ such that*

$$\sum_{k,\ell=1}^{d}\left\|\frac{\partial u_k}{\partial x_\ell}\right\|_{L^2(\Omega)}^2 \leq CS(u),$$

for every $u\in V$.

As an immediate consequence, one has the following corollary.

Corollary 8 *Let $\Omega\subseteq\mathbb{R}^d$ be as in (4). Suppose that Ω_1 and Ω_2 are both Sobolev and Friedrichs sets. Then V is a Hilbert space endowed with the norm $\sqrt{S(u)}$.*

Now that the aspects of functional analysis have been formulated, we rigorously frame Picone's problem (5)–(6) in the context of those elements. We start by defining operators and suitable functional spaces which allow us to produce an existence result.

2.1.1 An Existence Result for Picone's Problem (5)–(6)

Let $\Omega\subseteq\mathbb{R}^d$ be as in (4). Let $F^i : W^{1,2}(\Omega;\mathbb{R}^d)\times W^{1,2}(\Omega;\mathbb{R}^d)\to\mathbb{R}$, for $i=1,2$, be defined as

$$F^1(u^1,v^1) := 2h_1S(u^1,v^1)+k_1(\operatorname{div}u^1,\operatorname{div}v^1)_{L^2(\Omega_1)},$$

$$F^2(u^2,v^2) := 2h_2S(u^2,v^2)+k_2(\operatorname{div}u^2,\operatorname{div}v^2)_{L^2(\Omega_2)},$$

and

$$F(u, v) := F^1(u^1, v^1) + F^2(u^2, v^2),$$

where h_i and k_i are the Lamé constants for Ω_i, for $i = 1, 2$, $u = (u^1, u^2)$ and $v = (v^1, v^2)$.

Let $R_i : L^2(\Gamma) \to L^2(\Gamma)$ be a bounded linear operator for $i = 1, 2$. Define

$$G^i(u^i, v^i) := F^i(u^i, v^i) + \big(R_i T(u^i), T(v^i)\big)_{L^2(\Gamma;\mathbb{R}^d)},$$

and

$$G(u, v) := G^1(u^1, v^1) + G^2(u^2, v^2). \tag{9}$$

It is easy to see that (9) is a bilinear continuous operator in $W^{1,2}(\Omega; \mathbb{R}^d) \times W^{1,2}(\Omega; \mathbb{R}^d)$. Moreover, if R_i are positive and small enough, the operator in (9) is elliptic in V.

Now that the operators are defined, we can set the functional spaces where to look for the solution U to (5) and verify the transmission conditions (6).

Define the space $\mathcal{H}$ is the space of functions $u \in W^{1,2}(\Omega; \mathbb{R}^d)$ such that $Lu \in L^{\frac{2d}{d+2}}(\Omega; \mathbb{R}^d)$. Also, define N as the space of functions $u \in V$ such that $Lu \in L^{\frac{2d}{d+2}}(\Omega; \mathbb{R}^d)$, and

$$\langle Lu, v\rangle_{(L^{2d/(d+2)}(\Omega;\mathbb{R}^d), L^{2^*}(\Omega;\mathbb{R}^d))} = F(u, v),$$

for every $v \in V$. We endow N with the norm given by

$$\|u\|_N := \|u\|_V + \|Lu\|_{L^{2d/(d+2)}(\Omega;\mathbb{R}^d)}.$$

In what follows, we state a pivotal result from [45].

Proposition 9 *Let $\Omega \subseteq \mathbb{R}^d$ be as in* (4). *Suppose that Ω_1 and Ω_2 are both Sobolev and Friedrichs sets. Suppose further that F is elliptic in V. Then L is an isomorphism between N and $L^{\frac{2d}{d+2}}(\Omega; \mathbb{R}^d)$.*

Briefly, the proof of Proposition 9 relies on two facts. First, we have that $(V, \sqrt{S(\cdot)})$ is a Hilbert space. Secondly, one notices that, for every $u \in V$, one can define a linear continuous operator $\tilde{F} : V \to V$ such that

$$F^2(u, v) = F^1\left(\tilde{F}(u), v\right), \quad \tilde{F}(u) \in V,$$

and eventually $\tilde{F}$ is a symmetric operator.

The proof of Theorem 5 is now an easy consequence of the previous proposition; one argues as follows. Start by finding $U \in \mathcal{H}$ such that, for a given $H \in L^{\frac{2d}{d+2}}(\Omega; \mathbb{R}^d)$ and $h \in \mathcal{H}$, we have

$$LU = H$$

and

$$U - h \in N. \tag{10}$$

Now, we give a precise formulation of (10) that will lead to the transmission condition stated at the beginning. Integrating by parts, we get

$$\langle L_1 u^1, v^1 \rangle = F(u^1, v^1) - \int_{\partial\Omega_1} t_1(u^1) \cdot v^1 \, d\mathcal{H}^{d-1}.$$

Similarly, one has the formula for Ω_2. Now, $u \in N$ if and only if $u \in V$ and

$$\int_{\partial\Omega_1} t_1(u^1) \cdot v^1 \, d\mathcal{H}^{d-1} + \int_{\partial\Omega_2} t_2(u^2) \cdot v^2 \, d\mathcal{H}^{d-1} = 0$$

holds true for every $v \in V$. The last condition is equivalent to

$$\int_{S_1} t_1(u^1) \cdot v^1 \, d\mathcal{H}^{d-1} = 0, \tag{11}$$

$$\int_{S_2} t_2(u^2) \cdot v^2 \, d\mathcal{H}^{d-1} = 0, \tag{12}$$

and

$$\int_{\Gamma} t_1(u^1) \cdot v^1 + t_2(u^2) \cdot v^2 \, d\mathcal{H}^{d-1} = 0. \tag{13}$$

Since $T(v^1) = T(v^2)$, i.e. $v^1 = v^2$ on Γ, (11), (12) and (13) are equivalent to

$$\begin{cases} u^1 = u^2 & \text{on } \Gamma \\ t_1(u^1) + t_2(u^2) = 0 & \text{on } \Gamma \end{cases}$$

and

$$\begin{cases} t_1(u^1) = 0 & \text{on } S_1 \\ t_2(u^2) = 0 & \text{on } S_2. \end{cases}$$

This analysis completes the proof. The next section focuses on some examples of fixed transmission problems for which a regularity theory is available.

2.2 Some Developments in Regularity Theory

Important advances in the regularity theory of fixed transmission problems are in [43]. In that paper, the authors derive global $W^{1,\infty}$-estimates and $C^{1,\alpha}$-regularity for solutions to divergence form elliptic equations with piecewise Hölder continuous coefficients. We proceed by introducing some notation and making the assumptions more precise.

Let $\Omega \subset \mathbb{R}^d$ be a bounded domain with $C^{1,\alpha}$-regular boundary, $\alpha \in (0, 1)$, and let Ω_m, $1 \leq m \leq L$, be a finite number of disjoint subdomains of Ω, each with $C^{1,\alpha}$ boundary. Furthermore, suppose that

$$\overline{\Omega} = \bigcup_{m=1}^{L} \overline{\Omega_m}.$$

Given $\overline{x} \in \overline{\Omega}$, one supposes that there exists $r > 0$ and an appropriate rotation of our fixed coordinate system, such that the set

$$\left(\bigcup_{m=1}^{L} \partial\Omega_m \right) \cap B_r(\overline{x})$$

consists of the graphs of a finite number of $C^{1,\alpha}$-regular functions. Denote with $\ell(\overline{x}, r)$ the number of these functions, and let $K(\overline{x}, r)$ denote the maximum of their $C^{1,\alpha}$ norms. The quantity

$$\mathcal{K} := \sup_{\overline{x} \in \Omega} \inf_{r>0} \left(K(\overline{x}, r) + \ell(\overline{x}, r) + \frac{1}{r} \right)$$

is referred to as the $C^{1,\alpha}$ modulus of the total boundary set $(\cup_{m=1}^{L} \partial\Omega_m)$.

For $\mu \in (0, 1)$, let $A^{(m)} \in C^{\mu}(\overline{\Omega}_m)$ be a symmetric, positive definite matrix-valued function, and define

$$A(x) := A^{(m)}(x) \quad \text{in} \quad x \in \Omega_m, \; 1 \leq m \leq L. \tag{14}$$

Denote with $0 < \overline{\lambda} < \overline{\Lambda} < \infty$ the ellipticity constants associated with A. Similarly, let $g^{(m)} \in C^{0,\mu}(\overline{\Omega}; \mathbb{R}^d)$, and define

$$g(x) := g^{(m)}(x) \quad \text{in} \quad x \in \Omega_m, \; 1 \leq m \leq L. \tag{15}$$

Finally, suppose

$$h \in L^{\infty}(\Omega) \tag{16}$$

and

$$\varphi \in C^{1,\mu}(\partial\Omega). \tag{17}$$

The first of the results in [43] concerns $C^{1,\alpha'}$ interior estimates. In brief, this result asserts that the restriction of the solution u to each subdomain Ω_m can be extended to $\Omega_\varepsilon := \{x \in \Omega : \operatorname{dist}(x, \partial\Omega) > \varepsilon\}$ as a $C^{1,\alpha'}$ function, with a norm that is independent of the distances between the subdomain interfaces.

Theorem 10 *Suppose* (14)–(16) *hold. Suppose* α' *satisfies* $0 < \alpha' \leq \mu$ *and* $\alpha' < \alpha/((\alpha+1)d)$, *and let* $\varepsilon > 0$. *Then there exists* $C > 0$, *depending only on* $\Omega, d, \alpha, \alpha', \varepsilon, \overline{\lambda}, \overline{\Lambda}, \left\|A^{(m)}\right\|_{C^{0,\alpha'}(\overline{\Omega})}$, *and* $\mathcal{K}$, *such that, if* $u \in W^{1,2}(\Omega)$ *is a solution to*

$$\partial_i(A_{ij}\partial_j u) = h + \partial_i g_i \quad \textit{in} \quad x \in \Omega,$$

then

$$\max_{1\leq m\leq L} \|u\|_{C^{1,\alpha'}(\overline{\Omega}_m\cap\Omega_\varepsilon)} \leq C\Big(\|u\|_{L^\infty(\Omega)} + \|h\|_{L^\infty(\Omega)} + \max_{1\leq m\leq L}\left\|g^{(m)}\right\|_{C^{0,\alpha'}(\overline{\Omega})}\Big).$$

The variant of Theorem 10 covering boundary estimates is the following.

Theorem 11 *Suppose* (14)–(17) *hold. Suppose* α' *satisfies* $0 < \alpha' \leq \mu$ *and* $\alpha' < \alpha/((\alpha+1)d)$, *and let* $\varepsilon, r > 0$. *Then there exists* $C > 0$, *depending only on* $\Omega, d, \alpha, \alpha', \varepsilon, \overline{\lambda}, \overline{\Lambda}, \left\|A^{(m)}\right\|_{C^{0,\alpha'}(\overline{\Omega})}$ *and* $\mathcal{K}$, *such that, if for some* $\overline{x} \in \partial\Omega$, $u \in W^{1,2}(\Omega \cap B_{2r}(\overline{x}))$ *is a solution to*

$$\begin{cases} \partial_i(A_{ij}\partial_j u) = h + \partial_i g_i & \textit{in} \quad x \in \Omega \cap B_{2r}(\overline{x}) \\ u = \varphi & \textit{on} \quad x \in \partial\Omega \cap B_{2r}(\overline{x}), \end{cases}$$

then

$$\begin{aligned} \max_{1\leq m\leq L} \|u\|_{C^{1,\alpha'}(\overline{\Omega}_m\cap B_r(\overline{x}))} \leq C\Big(&\|u\|_{L^\infty(\Omega\cap B_r(\overline{x}))} + \|\varphi\|_{C^{1,\alpha'}(\partial\Omega\cap B_{2r}(\overline{x}))} \\ &+ \|h\|_{L^\infty(\Omega\cap B_r(\overline{x}))} \\ &+ \max_{1\leq m\leq L}\left\|g^{(m)}\right\|_{C^{0,\alpha'}(\overline{\Omega}\cap B_r(\overline{x}))}\Big). \end{aligned}$$

Combining the above interior and boundary estimates with the maximum principle (see [34]), we arrive at the following $C^{1,\alpha'}$ global estimate.

Corollary 12 *Suppose* (14)–(17) *hold. Suppose* α' *satisfies* $0 < \alpha' \leq \mu$ *and* $\alpha' < \alpha/((\alpha+1)d)$. *Then there exists* $C > 0$, *depending only on* $\Omega, d, \alpha, \alpha', \lambda, \overline{\Lambda}, \mathcal{K}$, *and* $\left\|A^{(m)}\right\|_{C^{0,\alpha'}(\overline{\Omega})}$, *such that, if* $u \in W^{1,2}(\Omega)$ *is a solution to*

$$\begin{cases} \partial_i(A_{ij}\partial_j u) = h + \partial_i g_i & \textit{in} \quad x \in \Omega \\ u = \varphi & \textit{on} \quad x \in \partial\Omega, \end{cases}$$

then

$$\max_{1\le m\le L}\|u\|_{C^{1,\alpha'}(\overline{\Omega}_m)}\le C\Big(\|\varphi\|_{C^{1,\alpha'}(\partial\Omega}+\|h\|_{L^\infty(\Omega)}$$
$$+\max_{1\le m\le L}\left\|g^{(m)}\right\|_{C^{0,\alpha'}(\overline{\Omega})}\Big).$$

We now describe the methods of the proof in [43]. To keep it brief and as clear as possible, we restrict our attention to Theorem 10 in the case $h=0$ and $g=0$. Let $\mathcal{A}(\overline{\lambda},\overline{\Lambda})$ denote the set of measurable, symmetric, positive definite matrix-functions $A(x)$ satisfying

$$\overline{\lambda}Id\le A(x)\le\overline{\Lambda}Id.$$

Define a scaling invariant subclass of $\mathcal{A}(\overline{\lambda},\overline{\Lambda})$, denoted by $\overline{\mathcal{A}}(\overline{\lambda},\overline{\Lambda})$, as follows. Fix $\ell\in\mathbb{N}$ and denote with $\{L_1,\dots,L_\ell\}$ a collection of ℓ parallel hyperplanes in $\mathbb{R}^d$, dividing $\mathbb{R}^d$ in $\ell+1$ regions. Denote such regions with $R_1,\dots,R_{\ell+1}$. Let $\overline{A}^{(1)},\dots,\overline{A}^{(\ell+1)}$ be any $\ell+1$ symmetric, positive definite constant matrices in $\mathcal{A}(\overline{\lambda},\overline{\Lambda})$ and define

$$\overline{A}_{ij}(x):=\overline{A}_{ij}^m\quad\text{on}\quad x\in R_m,\ 1\le m\le l+1.$$

The subclass $\overline{\mathcal{A}}(\overline{\lambda},\overline{\Lambda})$ consists of all such matrix functions $\overline{A}$.

The classical Schauder estimates, Cordes-Nirenberg estimates, and $W^{2,p}$ estimates can be viewed as perturbation theories from the corresponding estimates for solutions to the Laplace equation. Here, the approach is analogous, as the problem of interest is regarded as a perturbation of

$$\partial_i\left(\overline{A}_{ij}(x)\partial_j v\right)=0,\tag{18}$$

with $\overline{A}\in\overline{\mathcal{A}}(\overline{\lambda},\overline{\Lambda})$.

To establish Theorem 10 in the case $h=0$ and $g=0$, one first studies elliptic regularity estimates for solutions to (18). Although the hyperplanes in the definition of $\overline{\mathcal{A}}(\overline{\lambda},\overline{\Lambda})$ are allowed to get arbitrarily close to each other, estimates that are uniform in $\overline{A}\in\overline{\mathcal{A}}(\overline{\lambda},\overline{\Lambda})$ are available. Hence, the Caccioppoli inequality and the interior De Giorgi-Nash estimates yield bounds for all derivatives of the solution v in each region $\overline{R}_m$. More precisely, with $\Omega=(-1,1)^d$, one concludes that, for any positive integer k, any $\varepsilon>0$, any $\overline{A}\in\overline{\mathcal{A}}(\overline{\lambda},\overline{\Lambda})$ and any solution v to (18), it holds

$$\max_{1\le m\le l+1}\|v\|_{C^k(\overline{R}_m\cap(1-\varepsilon)\Omega)}\le C\|v\|_{L^\infty(\Omega)}.\tag{19}$$

Starting from (19), perturbation methods allow us to show that, for any $q>d$ and $\overline{\alpha}\in(0,1)$, there exists $\varepsilon_0>0$, depending only on $d,q,\overline{\alpha},\overline{\lambda}$ and $\overline{\Lambda}$, such that if $A\in\mathcal{A}(\overline{\lambda},\overline{\Lambda})$ and $\overline{A}\in\overline{\mathcal{A}}(\overline{\lambda},\overline{\Lambda})$ satisfy

$$\sup_{r\in(0,1)}r^{-\overline{\alpha}}\left(\frac{1}{|r\Omega|}\int_{r\Omega}\left|A(x)-\overline{A}(x)\right|^q\,\mathrm{d}x\right)^{\frac{1}{q}}\le\varepsilon_0\tag{20}$$

and $u \in W^{1,2}(\Omega)$ is a solution to

$$\partial_i(A_{ij}(x)\partial_j)u = 0 \text{ in } \Omega,$$

with

$$\|u\|_{L^\infty(\Omega)} \leq 1,$$

then there exists a continuous piecewise linear solution p to

$$\partial_i\left(\overline{A}_{ij}(x)\partial_j p\right) = 0 \text{ in } \frac{1}{4}\Omega,$$

whose coefficients are bounded in absolute value by a constant C, depending only on $d, q, \overline{\alpha}, \overline{\lambda}$, and $\overline{\Lambda}$, such that

$$|u(x) - p(x)| \leq C|x|^{1+\overline{\alpha}} \text{ in } \frac{1}{4}\Omega.$$

Under the hypotheses of Theorem 10, the condition (20) can be verified at every point $\overline{x} \in \Omega$ by translation and dilation, and by appropriate selection of $\overline{A_{\overline{x}}} \in \overline{\mathcal{A}}(\overline{\lambda}, \overline{\Lambda})$. The L^∞-interior estimates for the gradient of solutions to the equation $\partial_i(A_{ij}\partial_j u) = 0$ follow immediately. The Hölder interior estimates for the gradient require some further work, since at different points $\overline{x} \in \Omega$, the orientation of the hyperplanes associated with the matrices $\overline{A_{\overline{x}}}(x)$ differ by a rotation, determined by the geometry of Ω_m, and since $p_{\overline{x}}$ is only piecewise linear given a fixed set of planes.

2.3 *Geometry of the Interface and Optimal Regularity*

So far, we have dealt with transmission problems with smooth fixed interfaces. One of the main novelties presented in [20] is that the interface has only $C^{1,\alpha}$-regularity. In what follows, we detail the setting of that paper and describe its main contributions.

Let $\Omega \subseteq \mathbb{R}^d$ be a bounded domain. Fix $\Omega_1 \Subset \Omega$, and suppose that $\Gamma := \partial\Omega_1$ is a $C^{1,\alpha}$ manifold for some $\alpha \in (0, 1)$. Set $\Omega_2 := \Omega \setminus \overline{\Omega_1}$. The authors in that paper consider the problem of finding a continuous function $u : \overline{\Omega} \to \mathbb{R}$ such that

$$\begin{cases} \Delta u_1 = 0 & \text{in } \Omega_1 \\ \Delta u_2 = 0 & \text{in } \Omega_2 \\ u_2 = 0 & \text{in } \partial\Omega \\ u_1 = u_2 & \text{on } \Gamma \\ (u_1)_\nu - (u_2)_\nu = g & \text{on } \Gamma \end{cases} \tag{21}$$

where $u_1 := u|_{\overline{\Omega_1}}$, $u_2 := u|_{\overline{\Omega_2}}$, $g \in C^{0,\alpha}(\Gamma)$, and ν is the unit normal vector on Γ that is interior to Ω_1. The last two equations in (21) are called *transmission conditions*. Note that if $g \equiv 0$, then u is a harmonic function in Ω. Hence, in order to have a meaningful transmission condition, we suppose that

$$g(x) \geq 0 \quad \text{for} \quad x \in \Gamma.$$

The main result in [20] is the following theorem.

Theorem 13 *There exists a unique solution u to (21). Moreover, $u_1 \in C^{1,\alpha}(\overline{\Omega_1})$, $u_2 \in C^{1,\alpha}(\overline{\Omega_2})$, and there exists $C = C(d, \alpha, \Gamma) > 0$ such that*

$$\|u_1\|_{C^{1,\alpha}(\overline{\Omega_1})} + \|u_2\|_{C^{1,\alpha}(\overline{\Omega_2})} \leq C\|g\|_{C^{0,\alpha}(\Gamma)}.$$

In addition, $u \in C^{0,\text{Log-Lip}}(\overline{\Omega})$.

The existence and the uniqueness easily follow from a representation formula of the solution via the Green function G associated with the Laplace operator in Ω

$$u(x) = \int_\Gamma G(x, y)g(y)\,\mathrm{d}\mathcal{H}^{d-1} \quad \text{for} \quad x \in \Omega. \tag{22}$$

Thanks to the well-known properties of the Green function, we achieve, using the above formula, that $u \in C^{0,\text{Log-Lip}}(\overline{\Omega})$. The main issue is the regularity of u_i up to the interface for $i = 1, 2$. To establish this fact, one resorts to a number of building blocks. We continue by describing them and start with a discussion about the *flat problem*.

For $a \in \mathbb{R}$, we denote

$$\begin{aligned}
x &= (x', x_d) \in \mathbb{R}^{d-1} \times \mathbb{R} \\
B_{r,a} &= B_r(0', a) \\
B^+_{r,a} &= B_r(0', a) \cap \{x_d > a\} \\
B^-_{r,a} &= B_r(0', a) \cap \{x_d < a\} \\
T_{r,a} &= \{x \in B_r(0', a) : x_d = a\} \\
T_a &= B_1 \cap \{x_d = a\} \\
T^+_a &= B_1 \cap \{x_d \geq a\} \\
T^-_a &= B_1 \cap \{x_d \leq a\}.
\end{aligned}$$

When $a = 0$, we write $T = T_0$ and $B^\pm_r = B^\pm_{r,0}$. In the context of flat interfaces, the analogous of Theorem 13 reads as follows.

Theorem 14 *Let $r > 0$ and $a \in \mathbb{R}$. Given $\alpha, \gamma \in (0, 1)$, let $g \in C^{0,\alpha}(T_{r,a})$ and $f \in C^{0,\gamma}(\overline{B_{r,a}})$. Then there exists a unique solution $v \in C^\infty(B_{r,a} \setminus T_{r,a}) \cap C^{0,\gamma}(\overline{B_{r,a}})$ to the flat transmission problem*

$$\begin{cases} \Delta v = g\, d\mathcal{H}^{d-1}\big|_{T_{r,a}} & \text{in } B_{r,a} \\ v = f & \text{in } \partial B_{r,a}. \end{cases}$$

Moreover, if $v^{\pm} = v\chi_{B^{\pm}_{r,a}}$, *then* $v \in C^{1,\alpha}(\overline{B^{\pm}_{r/2,a}})$ *and*

$$\|v^{\pm}\|_{C^{1,\alpha}(\overline{B^{\pm}_{r/2,a}})} \leq C\left(\|g\|_{C^{0,\alpha}(T_{r,a})} + \|f\|_{L^{\infty}(\partial B_{r,a})}\right),$$

where C *depends only on* d, α, *and* r. *If* $g \in C^{k-1,\alpha}(T_{r,a})$, $k \geq 1$, *then* $v \in C^{k,\alpha}(\overline{B^{\pm}_{r/2,a}})$ *and*

$$\|v^{\pm}\|_{C^{k,\alpha}(\overline{B^{\pm}_{r/2,a}})} \leq C\left(\|g\|_{C^{k-1,\alpha}(T_{r,a})} + \|f\|_{L^{\infty}(\partial B_{r,a})}\right).$$

Briefly, the proof of the above theorem relies on the construction of two Dirichlet-Neumann problems, one for each hemisphere. Then, the solution to the original problem is given by the sum of the solutions of the aforementioned two problems.

Corollary 15 *Given* $|a| < 1/4$, $c_0 > 0$ *and* $f \in C^{0,\gamma}(\overline{B_1})$, *with* $\gamma \in (0,1)$, *there exists a unique solution* $v \in C^{\infty}(B_1 \setminus T_a) \cap C^{0,\gamma}(\overline{B_1})$ *to*

$$\begin{cases} \Delta v = c_0\, d\mathcal{H}^{d-1} & \text{in } B_1 \\ v = f & \text{on } \partial B_1 \end{cases}$$

such that, for any $k \geq 1$,

$$\|v^{\pm}\|_{C^{k,\alpha}(\overline{B_{1/2}} \cap T_a^{\pm})} \leq C\left(c_0 + \|f\|_{L^{\infty}(\partial B_1)}\right),$$

where $C = C(d, \alpha, k) > 0$.

Before continuing, further notation is required. Fix $\varepsilon > 0$, and let $\Omega_{\varepsilon} = \{x \in \Omega : \operatorname{dist}(x, \partial\Omega) < \varepsilon\}$ and $\Gamma_{\varepsilon} = \{x \in \Omega : \operatorname{dist}(x, \Gamma) < \varepsilon\}$. Consider the average

$$u_{\varepsilon}(x) = \frac{1}{|B_{\varepsilon}(x)|}\int_{B_{\varepsilon}(x)} u\, dy \quad \text{for} \quad x \in \Omega_{\varepsilon}.$$

Proposition 16 (Properties of averages) *Let* u *be the distributional solution given by (22). The following properties hold*

(*i*) *if* $B_{\varepsilon}(x) \cap \Gamma = \emptyset$ *then* $u_{\varepsilon}(x) = u(x)$;
(*ii*) $u_{\varepsilon} \to u$ *locally uniformly in* Ω *as* $\varepsilon \to 0^{+}$;
(*iii*) *if* $g \in L^{\infty}(\Omega)$ *then* $g_{\varepsilon} \in C_c(\Gamma_{\varepsilon})$, *where*

$$g_{\varepsilon}(x) = \frac{1}{|B_{\varepsilon}|}\int_{\Gamma \cap B_{\varepsilon}(x)} g\, d\mathcal{H}^{d-1} \quad \text{for} \quad x \in \Gamma_{\varepsilon}.$$

Moreover, $\Delta u_{\varepsilon}(x) = g_{\varepsilon}(x)$, *for every* $x \in \Omega_{\varepsilon}$.

The proof of the above proposition is quite simple, and it is essentially based on well-known properties of harmonic functions, the Lebesgue Dominated Convergence Theorem, and a change of variables. The following result is instrumental in the analysis.

Lemma 17 *Let* Γ *be as in Theorem 18. Define* $M := 1 + 2\theta$ *and let* $x \in B_{1-M\varepsilon}$ *be such that* $\mathrm{dist}(x, \Gamma) < \varepsilon$. *Then*

$$\begin{aligned}\{y' : (y', \psi(y')) \in B_\varepsilon(x)\} &\subseteq B'_{((M\varepsilon)^2-(x_d+\theta\varepsilon)^2)^{1/2}}(x') \\ &= \{y' : (y', -\theta\varepsilon) \in B_{M\varepsilon}(x)\}\end{aligned}$$

and

$$\begin{aligned}\{y' : (y', \psi(y')) \in B_{M\varepsilon}(x)\} &\supseteq B'_{(\varepsilon^2-(x_d+\theta\varepsilon)^2)^{1/2}}(x') \\ &= \{y' : (y', -\theta\varepsilon) \in B_\varepsilon(x)\}.\end{aligned}$$

Theorem 18 (Stability) *Let* $\varepsilon > 0, \theta < 1/2$ *and* $0 < \delta, \gamma < 1$ *be given. Let* $\Gamma = \{(y', \psi(y')) : y' \in B'_1\}$, *where* ψ *is a Lipschitz function. Suppose* Γ *is* $\theta\varepsilon$*-flat in* B_1, *in the sense that*

$$\Gamma \subseteq \{x \in B_1 : |x_d| < \theta\varepsilon\},$$

and that Γ *is also* ε*-horizontal in* B_1. *That is*

$$1 - \varepsilon \le \nu(x) \cdot (0', 1) = \left|1 + |D'\psi(x')|^2\right|^{-\frac{1}{2}} \le 1,$$

for every $x \in \Gamma$, *where* $\nu(x)$ *denotes the upward pointing normal on* Γ. *Then there exists* $C = C(d) > 0$ *such that, for any* $u \in C^{0,\gamma}(\overline{B_1})$ *and* $g \in L^\infty(\Gamma)$ *satisfying*

$$\begin{cases} \Delta u = g\, \mathrm{d}\mathcal{H}^{d-1}\big|_\Gamma & \text{in } B_1 \\ |g - 1| \le \delta & \text{on } \Gamma, \end{cases}$$

the classical solution $v \in C^\infty(B_1 \setminus T_{-\theta\varepsilon}) \cap C^{0,\gamma}(\overline{B_1})$ *to the flat problem*

$$\begin{cases} \Delta v = \mathrm{d}\mathcal{H}^{d-1}\big|_{T_{-\theta\varepsilon}} & \text{in } B_1 \\ v = u & \text{on } \partial B_1 \end{cases}$$

satisfies

$$|u - v| \le C\,(\theta + \delta + \varepsilon^\gamma) \quad \text{in } B_{1/2}.$$

Remark 19 The interface for the flat problem in Theorem 18 is $T_{-\theta\varepsilon} = B_1 \cap \{x_d = -\theta\varepsilon\}$, which lies below Γ in the x_d-direction. To approximate u with the solution to a flat problem where the interface lies above Γ in the x_d-direction, it is enough to consider the classical solution v to

$$\begin{cases} \Delta v = \mathrm{d}\mathcal{H}^{d-1}\big|_{T_{\theta\varepsilon}} & \text{in } B_1 \\ v = u & \text{on } \partial B_1. \end{cases}$$

In this case, the same conclusion as in Theorem 18 holds.

The proof of the stability result is quite technical, and it is obtained via a novel geometric approach, which is based on the mean value property and the maximum principle. What the stability tells us is that if the flatness and oscillation of the interface Γ are controlled, then one can construct a solution for a flat interface problem where the flat interface does not intersect Γ. Also, it is possible to quantify how close solutions are, depending only on the geometric properties of Γ and the basic regularity of u.

Now, suppose that Γ is an interface in B_1 given by the graph of a function $x_d = \psi(x') : T \to \mathbb{R}$. Thus, we can write $B_1 = \Omega_1 \cup \Gamma \cup \Omega_2$, where $\Omega_1 = \{x = (x', x_d) \in B_1 : x_d > \psi(x')\}$. It is also natural to suppose that $0 \in \Gamma$.

Lemma 20 *Given $0 < \alpha, \gamma < 1$, there exist $C_0 > 0$, $\lambda \in (0, 1/2)$, $0 < \theta, \delta, \varepsilon < \lambda$, depending only on d, α and γ, such that, for every $u \in C^{0,\gamma}(\overline{B_1})$ satisfying*

$$\begin{cases} \Delta u = g\, \mathrm{d}\mathcal{H}^{d-1}\big|_{\Gamma} & \textit{in } B_1 \\ |u| \leq 1 & \textit{in } B_1 \\ |g - 1| \leq \delta & \textit{on } \Gamma, \end{cases}$$

if Γ is $\theta\varepsilon$-flat and ε-horizontal in B_1, then there are linear polynomials $P_1(x) = A \cdot x + B$ and $Q_1(x) = C \cdot x + B$, with $A, C \in \mathbb{R}^d$, $B \in \mathbb{R}$ and $|A| + |B| + |C| \leq C_0$, for which

$$|u_1(x) - P_1(x)| \leq \lambda^{1+\alpha} \quad \textit{for} \quad x \in \Omega_1 \cap B_\lambda$$

and

$$|u_2(x) - Q_1(x)| \leq \lambda^{1+\alpha} \quad \textit{for} \quad x \in \Omega_2 \cap B_\lambda.$$

Moreover, $D'P_1 = D'Q_1'$ and $(P_1)_{x_d} - (Q_1)_{x_d} = 1$.

Lemma 21 *Given $\alpha \in (0, 1)$, there exist $C_0 > 0$, $\lambda \in (0, 1/2)$, $\delta \in (0, 1)$ depending only on d and α, such that, for any distributional solution $u \in C(\overline{B_1})$ to*

$$\begin{cases} \Delta u = g\, \mathrm{d}\mathcal{H}^{d-1}\big|_{\Gamma} & \textit{in } B_1 \\ |u| \leq 1 & \textit{in } B_1 \\ |g| \leq \delta & \textit{on } \Gamma, \end{cases}$$

there is a linear polynomial $P(x) = A \cdot x + B$, with $A \in \mathbb{R}^d$ and $B \in \mathbb{R}$ and $|A| + |B| \leq C_0$, satisfying

$$|u(x) - P(x)| \leq \lambda^{1+\alpha} \quad \textit{for} \quad x \in B_\lambda.$$

The previous two lemmas are key tools in the following theorem. In fact, the proof of the next result is based on a clever induction argument that involves a sequence of scaled fixed transmission problems. The very general idea is that flat solutions are asymptotically close to non-flat solutions. The theorem reads as follows.

Theorem 22 (Pointwise $C^{1,\alpha}$ boundary regularity) *Let*

$$\Gamma = \{(y', \psi(y')) : y' \in B_1'\},$$

where ψ is a $C^{1,\alpha}$-regular function, for some $\alpha \in (0,1)$. Suppose that $0 \in \Gamma$. Let u be a distributional solution to the transmission problem

$$\Delta u = g \, d\mathcal{H}^{d-1}|_{\Gamma},$$

where $g \in L^\infty(\Gamma) \cap C^{0,\alpha}(0)$ is nonnegative. Then there are linear polynomials $P(x) = A \cdot x + B$ and $Q(x) = C \cdot x + B$ such that

$$\begin{aligned} |u_1(x) - P(x)| &\leq D|x|^{1+\alpha} \quad \text{for} \quad x \in \Omega_1 \cap B_{1/2}, \\ |u_2(x) - Q(x)| &\leq D|x|^{1+\alpha} \quad \text{for} \quad x \in \Omega_2 \cap B_{1/2}, \end{aligned}$$

with

$$|A| + |B| + |C| + D \leq C_0 \|\psi\|_{C^{1,\alpha}(B_1')} \left([g]_{C^{0,\alpha}(0)} + \|g\|_{L^\infty(\Gamma)}\right)$$

and $C_0 = C_0(d, \alpha) > 0$.

To prove Theorem 13, one resort to Campanato's characterization of $C^{1,\alpha}$-spaces and a technical result that patches the interior and boundary estimates together.

Theorem 23 (Campanato's characterization of $C^{1,\alpha}$-spaces) *Let u be a measurable function defined on a bounded $C^{1,\alpha}$ domain Ω. Then $u \in C^{1,\alpha}(\overline{\Omega})$ if and only if there exists $C_0 > 0$ such that, for every $x \in \overline{\Omega}$, one finds a linear polynomial $Q_x(z)$ satisfying*

$$|u(z) - Q_x(z)| \leq C_0 |z - x|^{1+\alpha},$$

for every $z \in B_1(x) \cap \Omega$. In this case, if C_* denotes the least constant C_0 for which the property above holds, we have

$$\|u\|_{C^{1,\alpha}(\overline{\Omega})} \sim C_* + \sup_{x \in \overline{\Omega}} |Q_x|,$$

where $|Q_x|$ denotes the sum of the coefficients of the polynomial $Q_x(z)$.

At this point, the proof of Theorem 13 proceeds as follows. We first introduce three conditions and relate them with the conclusion of Theorem 13 through a proposition. In fact, the proposition ensures that if u_1 and u_2 satisfy the aforementioned conditions, then the regularity up to the fixed interface is available. We continue by introducing the conditions of interest.

From now on, let S be a collection of measurable functions defined on a bounded $C^{1,\alpha}$-regular domain $U \subset \mathbb{R}^d$.

Condition 24 (Interior estimates) *Let $u \in S$ and $d_x := \operatorname{dist}(x, \partial\Omega)$. There exist $A, C, D > 0$ such that, for every $x \in U$, one can find a linear polynomial $P_x(z)$ satisfying*

$$\|P_x\|_{L^\infty(B)} + d_x\|DP_x\|_{L^\infty(B)} \leq C\|u\|_{L^\infty(B)},$$

and

$$|u(z) - P_x(z)| \leq \left(A\frac{\|u\|_{L^\infty(B)}}{d_x^{1+\alpha}} + D\right)|z - x|^{1+\alpha},$$

for every $z \in B := B_{d_x/2}(x) \subset U$.

Condition 25 (Boundary estimates)*Let $u \in S$. There exists $E > 0$ such that, for every $y \in \partial U$, there is a linear polynomial $P_y(z)$ satisfying*

$$\|P_y\|_{L^\infty(U)} + \|DP_y\|_{L^\infty(U)} \leq E$$

and

$$|u(z) - P_y(z)| \leq E|z - y|^{1+\alpha},$$

for every $z \in \overline{U}$.

Condition 26 (Invariance property) *For every $u \in S$, and for every $y \in \partial U$ with a corresponding linear polynomial P_y as in Condition 25, the function $v = u - P_y$ also satisfies the estimates in Condition 24.*

Proposition 27 *Let S be a collection of measurable functions defined on a bounded $C^{1,\alpha}$-regular domain U. For $x \in U$, we let $d_x = \operatorname{dist}(x, \partial U)$. Fix $u \in S$, and suppose Conditions 24–26 are in force. Then $S \subseteq C^{1,\alpha}(\overline{U})$, and there exists $C > 0$, depending only on A, C, D, E, such that*

$$\|u\|_{C^{1,\alpha}(\overline{U})} \leq C\|u\|_{L^\infty(U)}.$$

Now, we show that u_1 and u_2 verify Conditions 24–26. This fact builds upon Proposition 27 to produce the conclusion of Theorem 13, completing its proof.

Let $u \in C^{0,\text{Log-Lip}}(\overline{\Omega})$ be given by (22). We consider only $u_2 : \overline{\Omega_2} \to \mathbb{R}$, as a similar argument yields the result for $u_1 : \overline{\Omega_1} \to \mathbb{R}$ and start with Condition 24.

Fix $x \in \Omega_2$. Since u_2 is harmonic, it is smooth in Ω_2, so we can define

$$P_x(z) := u_2(x) + Du_2(x) \cdot (z - x).$$

Then, by classical interior estimates for harmonic functions,

$$\begin{aligned}\|P_x\|_{L^\infty(B)} + d_x\|DP_x\|_{L^\infty(B)} \leq & \|u_2\|_{L^\infty(B)} + d_x\|Du_2\|_{L^\infty(B)} \\ & + d_x\|Du_2\|_{L^\infty(B)} \\ \leq & \|u_2\|_{L^\infty(B)} + 2d\|u_2\|_{L^\infty(B)} \\ \leq & (1+2d)\|Du_2\|_{L^\infty(B)}.\end{aligned}$$

Moreover,

$$\begin{aligned}|u_2(z) - P_x(z)| \leq & \left\|D^2u_2\right\|_{L^\infty(B)} |z-x|^2 \\ \leq & d\frac{\|u_2\|_{L^\infty(B)}}{d_x^2}|z-x|^2 \\ \leq & 2^{\alpha-1}d\frac{\|u_2\|_{L^\infty(B)}}{d_x^{1+\alpha}}|z-x|^{1+\alpha}.\end{aligned}$$

Concerning Condition 25, consider $\partial\Omega_2 = \Gamma \cup \partial\Omega$. If $y \in \Gamma$, Theorem 22 yields the existence of a linear polynomial $P_y(z)$ such that

$$\|P_y\|_{L^\infty(\Omega_2)} + \|DP_y\|_{L^\infty(\Omega_2)} \leq E$$

and

$$|u_2(z) - P_y(z)| \leq E|z-y|^{1+\alpha},$$

for every $z \in \overline{\Omega_2}$, with $E \leq C_0\|\psi\|_{C^{1,\alpha}(B_1')}\|g\|_{C^{0,\alpha}(\Gamma)}$, and $C_0 = C_0(d,\alpha) > 0$. If $y \in \partial\Omega \in C^\infty$, then, by classical boundary regularity for harmonic functions, $u_2 \in C^{1,\alpha}(\overline{B \cap \Omega})$, with $B := B_r(y)$, for some $r > 0$ sufficiently small. By Theorem 23, there exists a linear polynomial $P_y(z)$ such that

$$|u_2(z) - P_y(z)| \leq C_0|z-y|^{1+\alpha},$$

for every $z \in \overline{\Omega_2}$, for some $C_0 = C_0(d,\alpha) > 0$. This fact ensures that Condition 25 is in force.

Finally, we address Condition 26. Fix $y \in \partial\Omega_2$, and let P_y be the corresponding linear polynomial from Condition 25. Clearly, the function $v = u_2 - P_y$ is harmonic in Ω_2, so it satisfies the interior estimates from Condition 24.

Therefore, by Proposition 27, we have $u_2 \in C^{1,\alpha}(\overline{\Omega_2})$, and there exists a constant $C > 0$, depending only on d, α, Γ, such that

$$\|u_2\|_{C^{1,\alpha}(\overline{\Omega_2})} \leq C\|g\|_{C^{0,\alpha}(\Gamma)}.$$

We close this section by discussing an alternative proof to Theorem 13; it appeared in [29] and consists in re-casting the transmission problem in (21) as a Dirichlet problem driven by a PDE with a piecewise $C^{0,\alpha}$-regular right-hand side. Indeed, consider the problem of finding $w \in W^{1,2}(\Omega)$ satisfying

$$\begin{cases} \Delta w = c & \text{in} \quad \Omega_1 \\ w_\nu = g & \text{on} \quad \partial\Omega_1, \end{cases} \tag{23}$$

with

$$\int_{\Omega_1} w \mathrm{d}x = 0,$$

where

$$c := \frac{1}{\mathcal{H}^{d-1}(\Gamma)} \int_\Gamma g \, \mathrm{d}\mathcal{H}^{d-1}.$$

The existence of solutions to (23) follows from Sobolev inequalities and the Lax-Milgram Theorem. Also, usual results in elliptic regularity theory yield the existence of $\alpha \in (0, 1)$ and $C > 0$ such that $w \in C^{1,\alpha}(\Omega_1)$ with

$$\|w\|_{C^{1,\alpha}(\Omega_1)} \leq C \, \|g\|_{C^\alpha(\Gamma)} \, .$$

Once the existence and regularity for the auxiliary function w are available, we notice the solution u to (21) satisfies

$$\begin{cases} \Delta u = -\mathrm{div}\left(\chi_{\Omega_1} Dw\right) + \chi_{\Omega_1} c & \text{in} \quad \Omega \\ u = 0 & \text{on} \quad \partial\Omega. \end{cases} \tag{24}$$

Once again, the Lax-Milgram Theorem ensures the existence of a weak solution to (24). Concerning the regularity of u, an application of [28, Corollary 2 and Remark 3] yields the result. We notice the approach in [29] accommodates more general elliptic operators in divergence form, as long as uniform ellipticity is available. Furthermore, it covers nonhomogeneous equations.

3 A Quasilinear Fixed Transmission Problem

In this section, we discuss an example of a quasilinear fixed transmission problem modelled after the p-Laplace operator. The gist of the section is to extend to a class of degenerate, variational models, some of the results in [20], namely the regularity across the fixed transmission interface. The following material is based on the findings reported in [6].

We consider the following problem. Let $\Omega \subset \mathbb{R}^d$ be a bounded domain and fix $\Omega_1 \Subset \Omega$. Define $\Omega_2 := \Omega \setminus \overline{\Omega_1}$ and $\Gamma := \partial\Omega_1$. Suppose that the interface Γ is a $(d-1)$-manifold of class C^1. For a function $u : \overline{\Omega} \to \mathbb{R}$, we set

$$u_1 := u\big|_{\overline{\Omega_1}} \qquad \text{and} \qquad u_2 := u\big|_{\overline{\Omega_2}}.$$

The normal derivatives of u_1 and u_2 on the interface are defined as

$$\frac{\partial u_i}{\partial \nu} := Du_i \cdot \nu, \quad i = 1, 2,$$

where ν stands for the unit normal vector to Γ inwards Ω_1.

The variational formulation of the problem relies on a functional driven by a nonlinear function $g : \mathbb{R}_0^+ \to \mathbb{R}_0^+$, whose properties will be stated later. The PDE counterpart of the model is the quasilinear degenerate transmission problem consisting of finding a function $u : \overline{\Omega} \to \mathbb{R}$ such that

$$\begin{cases} \operatorname{div}\left(\frac{g(|Du_1|)}{|Du_1|} Du_1\right) = 0 & \text{in} \quad \Omega_1 \\ \operatorname{div}\left(\frac{g(|Du_2|)}{|Du_2|} Du_2\right) = 0 & \text{in} \quad \Omega_2, \end{cases} \tag{25}$$

under the following boundary and interface conditions

$$\begin{cases} u = 0 & \text{on} \quad \partial\Omega \\ \frac{g(|Du_1|)}{|Du_1|}\frac{\partial u_1}{\partial \nu} - \frac{g(|Du_2|)}{|Du_2|}\frac{\partial u_2}{\partial \nu} = f & \text{on} \quad \Gamma, \end{cases} \tag{26}$$

where the nonlinear function f is given. The integrability properties of the latter will lead to two different regularity results for the solutions.

Regarding the nonlinear function $g \in C^1\left(\mathbb{R}_0^+\right)$, we suppose that

$$g_0 \le \frac{tg'(t)}{g(t)} \le g_1, \quad \forall t > 0, \tag{27}$$

for fixed constants $1 \le g_0 \le g_1$. In addition, we impose the monotonicity condition

$$\left(\frac{g(|\xi|)}{|\xi|}\xi - \frac{g(|\zeta|)}{|\zeta|}\zeta\right) \cdot (\xi - \zeta) \ge C|\xi - \zeta|^p, \quad \forall \xi, \zeta \in \mathbb{R}^d, \tag{28}$$

for $p > 2$, fixed though arbitrary, and $C > 0$.

The work-horse of the theory is the case $g(t) = t^{p-1}$, with $p > 2$, which turns (25) into degenerate p-Laplace equations. A distinct example of nonlinearity within the scope of (27)–(28) is

$$g(t) := t^{p-1} \ln{(a + t)^\alpha},$$

for $p > 2$, $a > 1$ and $\alpha > 0$.

An important ingredient in the analysis is the primitive of g,

$$G(t) = \int_0^t g(s)\, ds, \quad t \ge 0.$$

We notice that G is a Young function. That is, it is left-continuous and convex; we refer the reader to [38, Definition 3.2.1]. Being a Young function, G allows us to define a functional space. This is the subject of the next definition.

Definition 28 (*Orlicz-Sobolev space*) Let G be a Young function. The Orlicz-Sobolev space $W^{1,G}(\Omega)$ is the set of weakly differentiable functions $u \in W^{1,1}(\Omega)$ such that

$$\int_\Omega G\left(|u(x)|\right)\mathrm{d}x + \int_\Omega G\left(|Du(x)|\right)\mathrm{d}x < \infty.$$

Reasoning as in the usual setting, it is also possible to define $W_0^{1,G}(\Omega)$. Now we introduce a notion of solution to (25)–(26).

Definition 29 We say that $u \in W_0^{1,G}(\Omega)$ is a weak solution of (25)–(26) if

$$\int_\Omega \frac{g\left(|Du|\right)}{|Du|} Du \cdot Dv\,\mathrm{d}x = \int_\Gamma f v\,\mathrm{d}\mathcal{H}^{d-1}, \qquad \forall\, v \in W_0^{1,G}(\Omega). \tag{29}$$

Remark 30 An important consideration on Definition 29 concerns the fact that the integrals in (29) are well-defined. This is actually the case; let $u, v \in W^{1,G}(\Omega)$ and notice that

$$\int_\Omega \frac{g(|Du|)}{|Du|} Du \cdot Dv\,\mathrm{d}x < \infty.$$

Also $tg(t) \leq CG(t)$, for $t \geq 0$, because g is increasing. Finally, $G(t+s) \leq C\big(G(t) + G(s)\big)$, for $t, s \geq 0$. As a result,

$$\begin{aligned}
\left|\int_\Omega \frac{g(|Du|)}{|Du|} Du \cdot Dv\,\mathrm{d}x\right| &\leq \int_\Omega g(|Du|)|Dv|\,\mathrm{d}x \\
&\leq \int_\Omega g(|Du| + |Dv|)(|Du| + |Dv|)\,\mathrm{d}x \\
&\leq C \int_\Omega G(|Du| + |Dv|)\,\mathrm{d}x \\
&\leq C \int_\Omega G(|Du|)\,\mathrm{d}x + C \int_\Omega G(|Dv|)\,\mathrm{d}x \\
&< \infty.
\end{aligned}$$

Remark 31 (*On the connection between* $W_0^{1,G}$ **and** $W_0^{1,p}$) Let $u \in W_0^{1,G}(\Omega)$, and suppose that (28) holds true. Then we see that $u \in W_0^{1,p}(\Omega)$. Indeed, that inequality yields

$$\begin{aligned}
\int_\Omega |Du|^p\,\mathrm{d}x &\leq C \int_\Omega g(|Du|)|Du|\,\mathrm{d}x \\
&\leq C \int_\Omega G(|Du|)\,\mathrm{d}x.
\end{aligned}$$

Remark 32 (*Existence and uniqueness of solutions*) The existence of a unique weak solution to (25)–(26) follows from approximation and monotonicity methods (see [5]; see also [44]). An important characterization of the weak solutions to (25) concerns the functional $I : W_0^{1,G}(\Omega) \to \mathbb{R}$ given by

$$I(u) = \int_\Omega G(|Du|)\, \mathrm{d}x - \int_\Gamma f u \,\mathrm{d}\mathcal{H}^{d-1}.$$

Indeed, a weak solution to (25)–(26) is a global minimizer for I, whose first compactly supported variation yields (29).

Since our interest lies in the Hölder-types of moduli of continuity, we proceed by introducing functional spaces suitable to our analysis. These are the Campanato and Morrey spaces, which provide us with useful characterizations for the regularity estimates on (25)–(26)

Definition 33 (*Campanato spaces*) We denote by $L_C^{p,\lambda}(\Omega;\mathbb{R}^d)$, with $1 \le p < \infty$ and $\lambda \ge 0$, the space of functions $u \in L^p(\Omega;\mathbb{R}^d)$ such that

$$[u]^p_{L_C^{p,\lambda}(\Omega;\mathbb{R}^d)} = \sup_{x^0\in\Omega,\rho>0} \frac{1}{\rho^\lambda}\int_{\Omega\cap B(x^0,\rho)} |u-(u)_{\Omega\cap B(x^0,\rho)}|^p\, \mathrm{d}x < \infty.$$

Definition 34 (*Morrey spaces*) We denote by $L_M^{p,\lambda}(\Omega;\mathbb{R}^d)$, with $1 \le p < \infty$ and $\lambda \ge 0$, the space of functions $u \in L^p(\Omega;\mathbb{R}^d)$ such that

$$\|u\|^p_{L_M^{p,\lambda}(\Omega)} = \sup_{x^0\in\Omega,\rho>0} \frac{1}{\rho^\lambda}\int_{\Omega\cap B(x^0,\rho)} |u|^p\, \mathrm{d}x < \infty.$$

We close this section with two technical results. The first one relates a decay regime with a Hölder-type of inequality.

Lemma 35 *Fix $R_0 > 0$ and let $\phi : [0, R_0] \to [0,\infty)$ be a non-decreasing function. Suppose there exist constants $C_1, \alpha, \beta > 0$, and $C_2, \mu \ge 0$, with $\beta < \alpha$, satisfying*

$$\phi(r) \le C_1\left[\left(\frac{r}{R}\right)^\alpha + \mu\right]\phi(R) + C_2R^\beta,$$

for every $0 < r \le R \le R_0$. Then, for every $\sigma \le \beta$, there exists $\mu_0 = \mu_0(C_1,\alpha,\beta,\sigma)$ such that, if $\mu < \mu_0$, for every $0 < r \le R \le R_0$, we have

$$\phi(r) \le C_3\left(\frac{r}{R}\right)^\sigma\left(\phi(R) + C_2R^\sigma\right),$$

where $C_3 = C_3(C_1,\alpha,\beta,\sigma) > 0$. Moreover,

$$\phi(r) \le C_4 r^\sigma,$$

where $C_4 = C_4(C_2, C_3, R_0, \phi(R_0), \sigma)$.

Our second technical lemma relates solutions to the transmission problem with solutions to the homogeneous equation. Indeed, it provides a lower bound for the difference of their G-integrals in terms of p-norms.

Lemma 36 *Let $w \in W^{1,G}(B_R)$. Suppose (27)–(28) is in force. Suppose further $h \in W^{1,G}_w(B_R)$ is a weak solution to*

$$\operatorname{div}\left(\frac{g(|Dh|)}{|Dh|}Dh\right) = 0 \quad \text{in } B_R.$$

Then there exists $C > 0$ such that

$$\int_{B_R} G(|Dw|) - G(|Dh|)\,\mathrm{d}x \geq C\int_{B_R} |D(w-h)|^p\,\mathrm{d}x.$$

For a proof of Lemma 36 we refer the reader to [6].

3.1 *Regularity Estimates in* BMO*−Spaces*

In the sequel, we consider $f \in L^\infty(\Omega)$. Our goal is to prove BMO−regularity estimates for the gradient of solutions *across the fixed transmission interface*. The following proposition is instrumental in our analysis.

Proposition 37 *Let $h \in W^{1,G}(B_R)$ be a weak solution of*

$$\operatorname{div}\left(\frac{g(|Dh|)}{|Dh|}Dh\right) = 0 \quad \text{in } B_R.$$

Suppose (27)–(28) hold true. Then there exist $C > 0$ and $\alpha \in (0, 1)$ such that, for every $r \in (0, R]$, we obtain

$$\int_{B_r} |Dh - (Dh)_r|\,\mathrm{d}x \leq C\left(\frac{r}{R}\right)^{d+\alpha}\int_{B_R} |Dh - (Dh)_R|\,\mathrm{d}x.$$

For a proof of Proposition 37, we refer the reader to [5].

Proposition 38 *Let $w \in W^{1,G}(B_R)$, and suppose $h \in W^{1,G}(B_R)$ is a weak solution of*

$$\operatorname{div}\left(\frac{g(|Dh|)}{|Dh|}Dh\right) = 0 \quad \text{in } B_R.$$

Suppose (27)–(28) hold true. Then there exists $C > 0$ such that, for every $0 < r \leq R$, one has

$$\int_{B_r} |Dw - (Dw)_r|\,dx \leq C\left(\frac{r}{R}\right)^{d+\alpha} \int_{B_R} |Dw - (Dw)_R|\,dx + C \int_{B_R} |Dw - Dh|\,dx$$

for $\alpha \in (0, 1)$ *as in Proposition 37.*

The proof of Proposition 38 is technical in nature, and we choose to omit it here (see [6, Proposition 2] for details). The main result in [6] reads as follows.

Theorem 39 (Gradient regularity in BMO−spaces) *Let* $u \in W_0^{1,G}(\Omega)$ *be a weak solution for the transmission problem* (25)–(26). *Suppose* (27)–(28) *are in force. Then* $Du \in \mathrm{BMO}_{\mathrm{loc}}(\Omega)$. *In addition, for every* $\Omega' \Subset \Omega$,

$$\|Du\|_{\mathrm{BMO}(\Omega')} \leq C,$$

where $C = C(d, \|f\|_{L^\infty(\Gamma)}, \mathrm{diam}(\Omega), \mathrm{dist}(\Omega', \partial\Omega)) > 0$.

The proof relies on an L^1-distance for the gradient of the solution u and the g-harmonic function h agreeing with u in the Sobolev sense on the boundary of B_R. In fact, we aim at producing

$$\int_{B_R} |D(u - h)|\,dx \leq CR^d. \tag{30}$$

The former inequality follows from an involved combination of the minimality of u, properties of g-harmonic functions, Lemma 36 and standard results, such as the Trace Theorem and the Poincaré Inequality.

Then (30) builds upon Proposition 38 to produce

$$\int_{B_r} |Du - (Du)_r|\,dx \leq C\left(\frac{r}{R}\right)^{d+\alpha} \int_{B_R} |Du - (Du)_R|\,dx + CR^d,$$

for every $0 < r \leq R$. Finally, an application of Lemma 35, implies

$$\int_{B_r} |Du - (Du)_r|\,dx \leq Cr^d, \quad \forall r \in (0, R],$$

which completes the proof.

At this point one resorts to embedding results for borderline spaces to obtain a modulus of continuity for the solution u in $C^{0,\mathrm{Log-Lip}}$−spaces. See [24, Theorem 3]. In fact, we have the following corollary.

Corollary 40 (Log-Lipschitz continuity estimates) *Let* $u \in W_0^{1,G}(\Omega)$ *be a weak solution for* (25)–(26). *Suppose* (27)–(28) *hold true. Then* $u \in C^{0,\mathrm{Log-Lip}}_{\mathrm{loc}}(\Omega)$. *Also, for every* $\Omega' \Subset \Omega$,

$$\|u\|_{C^{0,\text{Log-Lip}}(\Omega')} \leq C\left(\|u\|_{L^\infty(\Omega)} + \|f\|_{L^\infty(\Gamma)}\right),$$

where $C = C(p, d, \text{diam}(\Omega), \text{dist}(\Omega', \partial\Omega)) > 0$.

To complete this section we mention the case of unbounded interface data. Suppose $f \in W^{1,p'+\varepsilon}(\Omega)$, where $\varepsilon > 0$ depends on p and the dimension d. Hence, it is possible to prove the following regularity result in, Hölder spaces, for the solutions to (25)–(26).

Theorem 41 *Let u be a weak solution to the interface problem* (25)–(26), *and suppose* (27)–(28) *are in force. Let* $2 < p < d$, *and* $\varepsilon > 0$ *be such that*

$$\varepsilon < d - \frac{p}{p-1}.$$

Suppose further that $f \in W^{1,p'+\varepsilon}(\Omega)$. *Then* $u \in L_C^{1,d\varepsilon/(p'+\varepsilon)}(\Omega)$

Remark 42 (*Local boundedness*) We note that arguing along the same ideas as those presented in [53], one can prove that solutions to (25)–(26) are locally bounded (at the level of the functions and their gradients). Although we omit this result and its proof in the present manuscript, the reader may find the details in [6, Section 3].

Remark 43 (*Potential estimates and the p-Laplace operator*) For the case $g(t) := t^{p-1}$, one can use [39, Corollary 1, item (C9)] to derive the conclusion of Theorem 39. Because the interface Γ is locally of class C^1, it follows from the inequality

$$\int_{B_r \cap \Gamma} f \mathrm{d}\mathcal{H}^{d-1} \leq C r^{d-1};$$

see, for instance, [47, Proposition 3.5].

4 Free Transmission Problems

A variant of fixed transmission problems concerns models whose interfaces are solution-dependent. In this scenario, the models appear in the context of free boundary problems. As a consequence, an additional structure arises as part of the unknown; namely, the free interface.

We present this class of problems in the context of fully nonlinear elliptic equations, which are intrinsically non-variational and rely on the notion of viscosity solutions. We start by introducing the problem.

4.1 A Fully Nonlinear Free Transmission Problem: the Setting

Let $\Omega \subset \mathbb{R}^d$ be a bounded domain and consider $u \in C(\Omega)$. We define $\Omega^{\pm}(u)$ as

$$\Omega^+(u) := \{x \in \Omega \mid u(x) > 0\}$$

and

$$\Omega^-(u) := \{x \in \Omega \mid u(x) < 0\}.$$

Now, let $0 < \lambda \leq \Lambda$ be fixed constants. We let $F_1, F_2 : S(d) \to \mathbb{R}$ be (λ, Λ)-elliptic operators. That is,

$$\lambda \|N\| \leq F_i(M + N) - F_i(M) \leq \Lambda \|N\|,$$

for $i \in \{1, 2\}$, and every symmetric matrix $M, N \in S(d)$, with $N \geq 0$. We are interested in the problem

$$\begin{cases} F_1(D^2u) = f & \text{in } \ \Omega^+(u) \\ F_2(D^2u) = f & \text{in } \ \Omega^-(u). \end{cases} \tag{31}$$

The (free) interface Γ_u in this case is the topological boundary of the set where $u \neq 0$. That is,

$$\Gamma_u := \partial \{u > 0\} \cup \partial \{u < 0\}.$$

We consider two classes of solutions to (31): viscosity solutions (both in the continuous and the L^p-senses) and L^p-strong solutions. For the specifics on those notions, we refer the reader to [16, 26]; see also [34, Chap. 17]. It is worth noticing that when considering L^p-strong solutions one derives information on the transmission condition from the Sobolev properties of the solutions.

Because an L^p-strong solution is a function in $W^{2,d}(B_1)$, we have $Du = 0$ almost everywhere in $\{u = 0\}$. Hence, it is natural to equip (31) with the condition

$$|Du| = 0 \qquad \text{on} \qquad \partial \{u > 0\} \cup \partial \{u < 0\}.$$

A number of difficulties arise in the analysis of (31). First, the existence of solutions is a non-trivial matter. Indeed, because the diffusion process is discontinuous with respect to the solutions, the dependence of the equation on u is unknown. In particular, there is no a priori reason to expect it to be monotone. Hence, the equation governing the free transmission problem may lack properness and, therefore, a comparison principle.

In addition, the regularity of the solutions can no longer rely on the geometry of the transmission interface since such an object is unknown. As a consequence, the approach developed in [54] does not yield information on (31). In the sequel, we present recent developments bypassing those difficulties and extending the existence and regularity program to (31).

4.2 *The Existence of Solutions*

We consider the Dirichlet problem

$$\begin{cases} F_1(D^2u)\chi_{\{u>0\}} + F_2(D^2u)\chi_{\{u<0\}} = f & \text{in } \left(\Omega^+(u) \cup \Omega^-(u)\right) \cap \Omega \\ u = g & \text{on } \partial\Omega, \end{cases} \tag{32}$$

where $\Omega \subset \mathbb{R}^d$ is a bounded domain satisfying a uniform exterior one condition, $f \in L^p(\Omega)$, for $p > p_0$, and $g \in C(\partial\Omega)$ is a given boundary condition. We recall that $d/2 \leq p_0 = p_0(\Lambda/\lambda, d)$ is the exponent such that (λ, Λ)-elliptic equations with right-hand side in L^p, with $p > p_0$, are entitled to the Aleksandrov-Bakelman-Pucci maximum principle.

Remark 44 (*Escauriaza exponent $d/2 \leq p_0 = p_0(\Lambda/\lambda, d)$*) The connection of the exponent p_0 with fully nonlinear elliptic equations of the form $F(D^2u, x) = f$ appears in the work of Luis Escauriaza [30]. In that paper, the author proves a Harnack inequality of the form

$$\sup_{x\in B_{r/2}} u(x) \leq C\left(\inf_{x\in B_{r/2}} u(x) + r^{2-d/q}\, \|f\|_{L^p(B_r)}\right), \tag{33}$$

provided $u \geq 0$ solves $F(D^2u, x) = f$ and $f \in L^p(B_1)$, for $p > p_0$. This inequality follows from the improved integrability of the Green's function associated with the linearization of F. Such improved integrability of the Green's function was established in the work of Eugene Fabes and Daniel Stroock [31]. In the derivation of (33), one notices that Escauriaza's exponent p_0 is the conjugate of the improved integrability available for the Green's function of a (λ, Λ)-elliptic operator.

We present the following theorem.

Theorem 45 (Existence of viscosity solutions) *Let $\Omega \subset \mathbb{R}^d$ be a bounded domain satisfying a uniform exterior cone condition. Let $F_i : S(d) \to \mathbb{R}$ be (λ, Λ)-elliptic operators, for $i \in \{1, 2\}$, and some fixed constants $0 < \lambda \leq \Lambda$. Suppose $g \in C(\partial\Omega)$ and $f \in L^p(\Omega)$, for some $p > p_0$. Then there exists an L^p-viscosity solution $u \in C(\overline{\Omega})$ to* (32).

Before detailing the proof of Theorem 45, we mention that further conditions on the operators F_1 and F_2 allow us to obtain qualitative information on the solution to (32) whose existence follows from the theorem. The first condition concerns the local proximity of the operators F_1 and F_2; we suppose there exist constants $K, \tau > 0$ such that

$$|F_1(M) - F_2(M)| \leq K + \tau \, \|M\| \,, \tag{34}$$

for every $M \in S(d)$. Condition (34) unlocks a $C^{1,\alpha}$-estimate for the L^p-viscosity solution in Theorem 45.

The second condition we impose on F_i yields the existence of L^p-strong solutions to (32). This is a convexity-type of assumption and reads as follows. Suppose there exists $L, \sigma > 0$, and a *convex* (λ, Λ)-elliptic operator $F : S(d) \to \mathbb{R}$ such that

$$|F_i(M) - F(M)| \leq L + \sigma \, \|M\| \,, \tag{35}$$

for every $M \in S(d)$.

Theorem 46 (Existence of $C^{1,\alpha}$-regular solutions) *Let $\Omega \subset \mathbb{R}^d$ be a bounded domain satisfying a uniform exterior cone condition. Let $F_i : S(d) \to \mathbb{R}$ be (λ, Λ)-elliptic operators, for $i \in \{1, 2\}$, and some fixed constants $0 < \lambda \leq \Lambda$. Suppose $g \in C(\partial\Omega)$ and $f \in L^p(\Omega)$, for some $p > d$. Suppose further that* (34) *also holds; let $\alpha \in (0, 1)$ satisfy*

$$\alpha < \alpha_0 \quad \text{and} \quad \alpha \leq 1 - \frac{d}{p},$$

where $\alpha_0 \in (0, 1)$ corresponds to the C^{1,α_0}-regularity available for the solutions to $G = 0$ for any (λ, Λ)-elliptic operator G. Then there exists $\beta_0 = \beta_0(d, p, \lambda, \Lambda, \alpha) > 0$ such that, if the parameter $\tau > 0$ in (34) *satisfies $\tau \leq \beta_0$, then $u \in C^{1,\alpha}_{\mathrm{loc}}(\Omega)$ and, for every $\Omega' \Subset \Omega$, we have*

$$\|u\|_{C^{1,\alpha}(\Omega')} \leq C \left(1 + |F_1(0)| + |F_2(0)| + \|f\|_{L^p(\Omega)} + \|g\|_{L^\infty(\partial\Omega)}\right),$$

where $C = C(\alpha, d, p, \lambda, \Lambda, K, \tau, \mathrm{diam}(\Omega), \mathrm{dist}(\Omega', \partial\Omega))$.

We notice Theorem 46 not only ensures the existence of an L^p-viscosity solution to the free transmission problem but also provides a regularity estimate for this object. Of course, this is not a regularity result in the sense it applies only to the solution whose existence stems from the theorem. Also, one should compare the restrictions in Theorem 46 with the considerations on the optimal regularity result in [57, Sect. 4].

If we require F_1 and F_2 to be close, in the sense of (34), but also to satisfy a proximity regime with respect to a convex (λ, Λ)-operator F as in (35), the conclusion of Theorem 46 improves substantially. In fact, instead of a $C^{1,\alpha}$-regular L^p-viscosity solution, it is possible to establish the existence of a solution in $W^{2,p}(\Omega)$. As a consequence, under (35), it is possible to establish the existence of L^p-strong solutions for the free transmission problem. This is the content of the next theorem.

Theorem 47 (Existence of L^p-strong solutions) *Let $\Omega \subset \mathbb{R}^d$ be a bounded domain satisfying a uniform exterior cone condition. Let $F_i : S(d) \to \mathbb{R}$ be (λ, Λ)-elliptic operators, for $i \in \{1, 2\}$, and some fixed constants $0 < \lambda \leq \Lambda$. Suppose $g \in C(\partial\Omega)$ and $f \in L^p(\Omega)$, for some $p > p_0$. There exists $\beta_0 = \beta_0(d, p, \lambda, \Lambda, \alpha) > 0$ such* *that, if the parameter* $\sigma > 0$ *in* (35) *satisfies* $\sigma \leq \beta_0$, *then* (32) *has an* L^p*-strong solution* $u \in W^{2,p}(\Omega) \cap C(\overline{\Omega})$. *In addition, for every* $\Omega' \Subset \Omega$,

$$\|u\|_{W^{2,p}(\Omega')} \leq C\left(1 + |F_1(0)| + |F_2(0)| + \|f\|_{L^p(\Omega)} + \|g\|_{L^\infty(\partial\Omega)}\right),$$

where $C = C(\alpha, d, p, \lambda, \Lambda, L, \sigma, \mathrm{diam}(\Omega), \mathrm{dist}(\Omega', \partial\Omega))$.

The main difficulty in establishing the existence of viscosity solutions for (32) stems from the dependence of the operator on the zeroth order term u. In fact, the lack of properness rules out standard formulations of the comparison principle. As a result, Perron's method is no longer available.

We comment on the near-convexity condition in (35). For simplicity, one could simply require F_1 and F_2 to be convex operators. However, an important endeavour in the literature concerns the analysis of regularity estimates at the level of second derivatives (weak or classical) for non-convex operators. Indeed, in [52], the authors establish regularity estimates in Sobolev spaces $W^{2,p}$ by imposing an *asymptotic* convexity condition on F. This is done by means of the recession profile associated with F, namely

$$F^*(M) := \lim_{\mu \to \infty} \frac{1}{\mu} F(\mu M).$$

Indeed, by imposing convexity-like conditions on F_1^* and F_2^*, one would expect to derive Sobolev regularity estimates for the transmission problem under analysis.

However, a two-parameters regularization of the PDE in (32) turns the free transmission problem into an equation holding in the entire domain Ω, with no explicit dependence on the solution u. At this level, the comparison principle is available, and one can prove the existence of global barriers. Perron's method yields the existence of viscosity solutions. Then one proceeds by applying a fixed-point argument at the level of the functional parameter. Secondly, one sends the remaining parameter to zero; stability of viscosity solutions recovers a solution to (32) and the proof is complete.

In what follows, we introduce some of the ingredients in those arguments. The main idea concerns the regularization of the equation in (32). Let $v \in C(\overline{\Omega})$ agree with the Dirichlet data g on the boundary. Fix $\varepsilon > 0$, arbitrary. Define a function $g_\varepsilon^v : \Omega \to \mathbb{R}$ as

$$g_\varepsilon^v(x) := \max\left(\min\left(\frac{v(x) + \varepsilon}{2\varepsilon}, 1\right), 0\right) \quad \text{in} \quad \Omega,$$

with $g_\varepsilon^v \equiv 0$ in $\mathbb{R}^d \setminus \Omega$. Notice that $g_\varepsilon^v(x) = 1$ in $\{v > \varepsilon\}$, whereas $g_\varepsilon^v(x) = 0$ in $\{v < -\varepsilon\}$. We consider the convolution of g_ε^v with a standard mollifying kernel η_ε to obtain

$$h_\varepsilon^v := g_\varepsilon^v * \eta_\varepsilon.$$

Once h_ε^ν is available, we introduce the auxiliary operator $G_\varepsilon^\nu : \Omega \times S(d) \to \mathbb{R}$ given by

$$G_\varepsilon^\nu(x, M) := h_\varepsilon^\nu(x)F_1(M) + (1 - h_\varepsilon^\nu(x))F_2(M).$$

A straightforward computation implies three properties of the operator G_ε^ν. First, because F_1 and F_2 are (λ, Λ)-elliptic, we conclude G_ε^ν is also a (λ, Λ)-elliptic operator. One can also prove the existence of a constant $K_\varepsilon^\nu > 0$ such that

$$\left|G_\varepsilon^\nu(x, M) - G_\varepsilon^\nu(y, M)\right| \le K_\varepsilon^\nu |x - y| \left(1 + \|M\|\right),$$

for every $x, y \in \Omega$ and every $M \in S(d)$. That is, we conclude that G_ε^ν is Lipschitz-continuous in the space-variable, locally uniformly in M.

A fundamental information on G_ε^ν concerns its connection with the conditions in (34) and (35). Indeed, the definition of G_ε^ν preserves those conditions, up to an adjustment in the constants. If (34) holds, we have

$$\left|G_\varepsilon^\nu(x, M) - G_\varepsilon^\nu(y, M)\right| \le 2\left(K + \tau \|M\|\right),$$

for every $x, y \in \Omega$ and every $M \in S(d)$. Also, if (35) is in force, we get

$$\left|G_\varepsilon^\nu(x, M) - F(M)\right| \le L + \sigma \|M\|,$$

for every $x \in \Omega$ and $M \in S(d)$. For a proof of those properties, we refer the reader to [51, Lemma 1].

Once G_ε^ν is well defined, we consider the Dirichlet problem

$$\begin{cases} G_\varepsilon^\nu(x, D^2 u_\varepsilon^\nu) = f & \text{in } \Omega \\ u_\varepsilon^\nu = g & \text{on } \partial\Omega. \end{cases} \tag{36}$$

In [51], the authors observe the comparison principle is available for the equation in (36). In addition, they establish the existence of global sub and supersolutions $\underline{u}$ and $\overline{u}$ and notice that such functions are independent of ν and $\varepsilon > 0$. Perron's method then implies the existence of an L^p-viscosity solution u_ε^ν to (36). Moreover, $\underline{u} \le u_\varepsilon^\nu \le \overline{u}$.

At this point, one considers two objects. First, a subset $B \subset C(\overline{\Omega})$ defined as

$$B := \left\{ v \in C(\overline{\Omega}) \mid \underline{u} \le v \le \overline{u} \right\}.$$

Then we introduce an operator T defined on B as follows: for $v \in B$, one solves (36) to obtain the unique solution u_ε^ν. One sets $Tv := u_\varepsilon^\nu$. These ingredients satisfy certain properties. In fact, B is a close and convex subset of $C(\overline{\Omega})$. Also, T maps B to itself and $T(B)$ is a precompact set in $C(\overline{\Omega})$. Finally, one notices that $T : B \to B$ is continuous.

As a consequence, an application of the Schauder Fixed Point Theorem yields the existence of an L^p-viscosity solution to

$$\begin{cases} G^u_\varepsilon(x, D^2 u_\varepsilon) = f & \text{in } \Omega \\ u_\varepsilon = g & \text{on } \partial\Omega. \end{cases}$$

Finally, to prove the existence of an L^p-viscosity solution to (32), one takes the limit $\varepsilon \to 0$ and resorts to the stability of viscosity solutions. These steps lead to the conclusion of Theorem 45. To establish Theorem 46 under (34), one can argue through well-understood arguments, such as in [18, Theorem 8.3] or [56, Theorem 2.1]. Concerning Theorem 47, one notices that (35) frames the problem in the context of classical $W^{2,p}$-regularity theory for fully nonlinear elliptic equations; see [18, Chap. 7].

Once we have discussed the existence of the solutions to (32), we proceed by addressing their regularity estimates. This is the subject of the next section.

4.3 *Regularity for Fully Nonlinear Free Transmission Problems*

Once the existence of L^p-viscosity and L^p-strong solutions has been addressed in the literature, the natural question concerns their regularity. This is the subject of [50], whose general lines we discuss in what follows.

In that work, the authors examine L^p-strong solutions to (31) and establish two classes of regularity estimates. Under a near-convexity condition as in (35), they show that strong solutions are locally of class $C^{1,\mathrm{Log-Lip}}$. Furthermore, under an additional (pointwise) condition on the density of the negative phase, they prove that solutions satisfy a (pointwise) quadratic growth regime.

Once again, the main difficulty in the analysis of (31) is the discontinuity of the operator with respect to the solution. Indeed, standard results such as the Harnack inequality or the maximum principle are not available. To bypass the lack of usual ideas and methods, the argument in [50] relates (31) with two viscosity inequalities, *holding in the entire domain*, for which the theory is available.

In fact, let $u \in W^{2,d}(\Omega)$ be an L^d-strong solution to (31). Hence, it satisfies

$$\min\left\{F_1(D^2u(x)), F_2(D^2u(x))\right\} \leq 1 \qquad \text{a.e. in } \Omega$$

and

$$\max\left\{F_1(D^2u(x)), F_2(D^2u(x))\right\} \geq -1 \qquad \text{a.e. in } \Omega.$$

One then explores the connection between L^d-strong and L^d-viscosity solutions, as detailed in [16]. It follows from [16, Lemma 2.5] that u is an L^d-viscosity solution to

$$\begin{cases} \min\left\{F_1(D^2u), F_2(D^2u)\right\} \leq 1 & \text{in } \Omega \\ \max\left\{F_1(D^2u), F_2(D^2u)\right\} \geq -1 & \text{in } \Omega. \end{cases} \tag{37}$$

We notice, however, the ingredients in (37) are continuous. Hence, u is indeed a C-viscosity solution to this pair of inequalities, and one immediately concludes that $u \in C^{\alpha}_{\text{loc}}(\Omega)$, with estimates. This fact yields compactness for any family of solutions to (37). Also, the operators governing these inequalities are entitled to stability results for viscosity solutions, in the spirit of [18, Proposition 4.11].

Theorem 48 (Local $C^{1,\text{Log-Lip}}$-regularity) *Let $u \in W^{2,d}_{\text{loc}}(B_1)$ be a strong solution to* (31)*. Suppose F_1 and F_2 are convex, (λ, Λ)-elliptic operators. Suppose further that* (35) *is in force. There exists $0 < \beta_0 = \beta_0(d, \lambda, \Lambda) \ll 1$ such that, if $\sigma \leq \beta_0$, we have $u \in C^{1,\text{Log-Lip}}_{\text{loc}}(B_1)$, and there exists $C > 0$ such that*

$$\sup_{x \in B_r(x_0)} |u(x) - u(x_0) - Du(x_0) \cdot (x - x_0)| \leq C r^2 \ln \frac{1}{r},$$

for every $x_0 \in \Omega$ and $r > 0$ satisfying $B_r(x_0) \Subset \Omega$. In addition, $C = C(d, \lambda, \Lambda, \|u\|_{L^\infty(\Omega)})$.

The proof of Theorem 48 follows from approximation methods, along with the ideas introduced in the work of Luis Caffarelli [17]; see also [15]. For a detailed argument, we refer the reader to [50]. We mention the analysis of borderline Log-Lipschitz regularity estimates in the context of fully nonlinear elliptic equations appeared for the first time in [57, Theorem 3].

An interesting question concerning the solutions of (31) arises from Theorem 48. Indeed, one may enquire into the further conditions on the problem that would switch the regularity regime from the borderline scenario $C^{1,\text{Log-Lip}}$ to that of $C^{1,1}$-regularity estimates.

A condition that allows us to improve the information in Theorem 48 regards the density of the negative phase in a vicinity of a given point $x_0 \in \Omega$ on the free boundary. We proceed by introducing some ingredients.

The free boundary associated with (31) is the set

$$\Gamma(u) := \left(\partial\Omega^+(u) \cup \partial\Omega^-(u)\right) \cap \Omega.$$

A point $x_0 \in \Gamma(u)$ may fall within three distinct categories. First, it can be a *one-phase point*, in the sense that it lies on the free boundary between the positive or the negative phase and the set $\{u = 0\}$. More rigorously, we say that $x_0 \in \Gamma(u)$ is a one-phase point if

$$x_0 \in \left(\partial\Omega^{\pm}(u) \setminus \partial\Omega^{\mp}(u)\right) \cap \Omega.$$

When examining a one-phase free boundary point, the regularity analysis amounts to the study of an obstacle problem, which has been well-understood and documented; see, for instance, [40].

Another situation refers to points $x_0 \in \Gamma(u)$ sitting on a portion of the free boundary separating the positive and the negative phases. These are called *two-phase points* and satisfy

$$x_0 \in \left(\partial\Omega^+(u) \cap \partial\Omega^-(u)\right) \cap \Omega.$$

The study of the regularity of the solutions or the free boundary properties around two-phase points is completely open in the context of (31).

However, among such points, we highlight $x^* \in \Gamma(u)$ for which

$$|B_r(x^*) \cap \{u = 0\}| > 0, \tag{38}$$

for every $0 < r \ll 1$. A two-phase point satisfying (38) is called a *branch point* for (31). We denote with $\Gamma_{\text{BR}}(u) \subset \Gamma(u)$ the set of branch points for the solution u.

A further ingredient in the analysis of quadratic growth for the solutions to (31) is the density of the negative phase. Given a point $x_0 \in \Gamma(u)$, we define $V_r(x_0, u)$ as

$$V_r(x_0, u) := \frac{\text{Vol}\left(B_r(x_0) \cap \Omega^-(u)\right)}{r^d};$$

the quantity $V_r(x_0, u)$ amounts to the density of the negative phase around a free boundary point and plays a critical role in the regularity theory for the solutions. Indeed, in [50], the authors suppose F_1 and F_2 to be convex (λ, Λ)-elliptic operators, positively homogeneous of degree one. In addition, if there exist a constant $C_0 > 0$ and $x_0 \in \Gamma(u)$ such that

$$V_r(x_0, u) \leq C_0,$$

they prove

$$\sup_{x \in B_r(x_0)} |u(x)| \leq Cr^2,$$

for every $0 < r \ll 1$, provided C_0 can be taken *arbitrarily small.*

Were the smallness condition available locally along $\Gamma(u)$, the quadratic growth would imply $C^{1,1}$-regularity estimates for the solutions to (31). However, such a condition would imply the negative phase is empty in such a neighbourhood, trivializing the problem. Instead of aiming at a local $C^{1,1}$-regularity result, the authors consider branch points $x^* \in \Gamma_{\text{BR}}(u)$ and prove the following theorem.

Theorem 49 (Quadratic growth at branch points) *Let $u \in W^{2,d}_{\text{loc}}(B_1)$ be a strong solution to* (31). *Suppose the operators F_1 and F_2 are convex, positively homogeneous of degree one, and (λ, Λ)-elliptic. Let $x^* \in \Gamma_{\text{BR}}(u)$ be such that $V_r(x_0, u) \leq C_0$. Then there exists $0 < \overline{C_0} \ll 1$ such that, if $C_0 \leq \overline{C_0}$, one can find a universal constant $C > 0$ for which*

$$\sup_{x \in B_r(x^*)} |u(x)| \leq Cr^2,$$

for every $0 < r \ll 1$.

We present the general lines of the proof of Theorem 49 in the sequel. It relies on a dyadic analysis combined with the maximum principle and a scaling argument. The latter uses the L^∞-norms of u as a normalization factor, introducing an additional dependence on this quantity into the estimates.

The strategy used in [50] is inspired by arguments first launched in [19] in the analysis of a free boundary arising in the Pompeiu problem. In that paper, the diffusion process is driven by the Laplace operator. The fully nonlinear counterpart of [19] appeared in [41]. Here, the authors identify that convexity and homogeneity of degree one are the precise conditions allowing to switch from the linear to the nonlinear setting.

To detail the proof of Theorem 49, we start by defining a subset of the natural numbers related to upper bounds for u in dyadic balls. Fix $x^* \in \Gamma_{\mathrm{BR}}(u) \cap \Omega$. The maximal subset of $\mathbb{N}$ whose elements j are such that

$$\sup_{x \in B_{2^{-j-1}}(x^*)} |u(x)| \geq \frac{1}{16} \sup_{x \in B_{2^{-j}}(x^*)} |u(x)|$$

is denoted with $\mathcal{M}(x^*, u)$.

Proposition 50 *Let $u \in W^{2,d}_{\mathrm{loc}}(B_1)$ be a strong solution to* (31). *Suppose the operators F_1 and F_2 are convex, positively homogeneous of degree one, and (λ, Λ)-elliptic. Let $x^* \in \Gamma_{\mathrm{BR}}(u)$ be such that*

$$V_{2^{-j}}(x^*, u) < C_0,$$

for every $j \in \mathcal{M}(x^*, u)$, where $C_0 \leq \overline{C_0}$. Then

$$\sup_{x \in B_{2^{-j}}(x^*)} |u(x)| \leq \frac{1}{C_0} 2^{-2j}, \qquad \forall j \in \mathcal{M}(x^*, u).$$

For a proof of Proposition 50, we refer the reader to [50, Proposition 3]. Once the quadratic growth holds in the set $\mathcal{M}(x^*, u)$, one extends this fact to the natural numbers $\mathbb{N}$. This is the subject of the next proposition.

Proposition 51 *Let $u \in W^{2,d}_{\mathrm{loc}}(B_1)$ be a strong solution to* (31). *Suppose the operators F_1 and F_2 are convex, positively homogeneous of degree one, and (λ, Λ)-elliptic. Let $x^* \in \Gamma_{\mathrm{BR}}(u)$ be such that*

$$V_{2^{-j}}(x^*, u) < C_0,$$

for every $j \in \mathcal{M}(x^*, u)$, where $C_0 \leq \overline{C_0}$. Then

$$\sup_{x \in B_{2^{-j}}(x^*)} |u(x)| \leq \frac{4}{C_0} 2^{-2j}, \qquad \forall j \in \mathbb{N}.$$

For the detail in the proof of Proposition 51, we refer the reader to [50, Proposition 4]. Finally, a discrete-to-continuous argument builds upon Proposition 51 to complete the proof of Theorem 49.

Given $0 < r \ll 1$, let $j \in \mathbb{N}$ be such that $2^{-(j+1)} \leq r < 2^{-j}$. Hence,

$$\sup_{B_r} |u(x)| \leq \sup_{B_{2^{-j}}} |u(x)| \leq C \left[\left(\frac{1}{2} \right)^{j+1-1} \right]^2 \leq Cr^2,$$

and the proof is complete.

Acknowledgements All authors are partially supported by the Centre for Mathematics of the University of Coimbra (UIDB/00324/2020, funded by the Portuguese Government through FCT/MCTES).

EP is partially supported by FAPERJ (grants E26/200.002/2018 and E26/201.390/2021).

JMU is partially supported by the King Abdullah University of Science and Technology (KAUST).

References

1. Amaral, M.D., Teixeira, E.V.: Free transmission problems. Comm. Math. Phys. **337**(3), 1465–1489 (2015). https://doi.org/10.1007/s00220-015-2290-3
2. Araújo, D.a.J., Ricarte, G., Teixeira, E.V.: Geometric gradient estimates for solutions to degenerate elliptic equations. Calc. Var. Part.L Differ. Equ. **53**(3–4), 605–625 (2015). https://doi.org/10.1007/s00526-014-0760-7
3. Bao, E.S., Li, Y.Y., Yin, B.: Gradient estimates for the perfect conductivity problem. Arch. Ration. Mech. Anal. **193**(1), 195–226 (2009). https://doi.org/10.1007/s00205-008-0159-8
4. Bao, E.S., Li, Y.Y., Yin, B.: Gradient estimates for the perfect and insulated conductivity problems with multiple inclusions. Comm. Part. Differ. Equ. **35**(11), 1982–2006 (2010). https://doi.org/10.1080/03605300903564000
5. Baroni, P.: Riesz potential estimates for a general class of quasilinear equations. Comm. Part. Differ. Equ. **53**(3–4), 803–846 (2015). https://doi.org/10.1007/s00526-014-0768-z
6. Bianca, V., Pimentel, E.A., Urbano, J.M.: BMO-regularity for a degenerate transmission problem. Anal. Math. Phys. **14**(1), Paper No. 9, 26 (2024). https://doi.org/10.1007/s13324-023-00867-x
7. Birindelli, I., Demengel, F.: Comparison principle and Liouville type results for singular fully nonlinear operators. Ann. Fac. Sci. Toulouse Math. (6) **13**(2), 261–287 (2004)
8. Birindelli, I., Demengel, F.: First eigenvalue and maximum principle for fully nonlinear singular operators. Adv. Differ. Equ. **11**(1), 91–119 (2006)
9. Birindelli, I., Demengel, F.: The Dirichlet problem for singular fully nonlinear operators. Discret. Contin. Dyn. Syst. 110–121 (2007)
10. Birindelli, I., Demengel, F.: Eigenvalue, maximum principle and regularity for fully non linear homogeneous operators. Commun. Pure Appl. Anal. **6**(2), 335–366 (2007). https://doi.org/10.3934/cpaa.2007.6.335
11. Birindelli, I., Demengel, F.: Eigenvalue and Dirichlet problem for fully-nonlinear operators in non-smooth domains. J. Math. Anal. Appl. **352**(2), 822–835 (2009). https://doi.org/10.1016/j.jmaa.2008.11.012
12. Birindelli, I., Demengel, F.: $\mathscr{C}^{1,\beta}$ regularity for Dirichlet problems associated to fully nonlinear degenerate elliptic equations. ESAIM Control Optim. Calc. Var. **20**(4), 1009–1024 (2014). https://doi.org/10.1051/cocv/2014005

13. Borsuk, M.: Transmission problems for elliptic second-order equations in non-smooth domains. Frontiers in Mathematics. Birkhäuser/Springer Basel AG, Basel (2010). https://doi.org/10.1007/978-3-0346-0477-2
14. Borsuk, M.V.: A priori estimates and solvability of second order quasilinear elliptic equations in a composite domain with nonlinear boundary condition and conjugacy condition. Trudy Mat. Inst. Steklov. **103**, 15–50. (loose errata) (1968)
15. Caffarelli, L.: Elliptic second order equations. Rend. Sem. Mat. Fis. Milano **58**, 253–284 (1988). https://doi.org/10.1007/BF02925245
16. Caffarelli, L., Crandall, M.G., Kocan, M., Swięch, A.: On viscosity solutions of fully nonlinear equations with measurable ingredients. Comm. Pure Appl. Math. **49**(4), 365–397 (1996). https://doi.org/10.1002/(SICI)1097-0312(199604)49:4<365::AID-CPA3>3.3.CO;2-V
17. Caffarelli, L.A.: Interior a priori estimates for solutions of fully nonlinear equations. Ann. of Math. (2) **130**(1), 189–213 (1989). https://doi.org/10.2307/1971480
18. Caffarelli, L.A., Cabré, X.: Fully nonlinear elliptic equations. In: American Mathematical Society Colloquium Publications, vol. 43. American Mathematical Society, Providence, RI (1995). https://doi.org/10.1090/coll/043
19. Caffarelli, L.A., Karp, L., Shahgholian, H.: Regularity of a free boundary with application to the Pompeiu problem. Ann. of Math. (2) **151**(1), 269–292 (2000). https://doi.org/10.2307/121117
20. Caffarelli, L.A., Soria-Carro, M., Stinga, P.R.: Regularity for $C^{1,\alpha}$ interface transmission problems. Arch. Ration. Mech. Anal. **240**(1), 265–294 (2021). https://doi.org/10.1007/s00205-021-01611-0
21. Campanato, S.: Sul problema di M. Picone relativo all'equilibrio di un corpo elastico incastrato. Ricerche Mat. **6**, 125–149 (1957)
22. Campanato, S.: Sui problemi al contorno per sistemi di equazioni differenziali lineari del tipo dell'elasticità. I. Ann. Scuola Norm. Sup. Pisa Cl. Sci. (3) **13**, 223–258 (1959)
23. Campanato, S.: Sui problemi al contorno per sistemi di equazioni differenziali lineari del tipo dell'elasticità. II. Ann. Scuola Norm. Sup. Pisa Cl. Sci. (3) **13**, 275–302 (1959)
24. Cianchi, A.: Continuity properties of functions from Orlicz-Sobolev spaces and embedding theorems. Ann. Scuola Norm. Sup. Pisa Cl. Sci. (4) **23**(3), 575–608 (1996)
25. Colombo, M., Kim, S., Shahgholian, H.: A transmission problem with (p, q)-Laplacian. Comm. Part. Differ. Equ. **48**(2), 315–349 (2023). https://doi.org/10.1080/03605302.2023.2175216
26. Crandall, M.G., Ishii, H., Lions, P.L.: User's guide to viscosity solutions of second order partial differential equations. Bull. Am. Math. Soc. (N.S.) **27**(1), 1–67 (1992). https://doi.org/10.1090/S0273-0979-1992-00266-5
27. De Filippis, C.: Fully nonlinear free transmission problems with nonhomogeneous degeneracies. Interfaces Free Bound **24**(2), 197–233 (2022). https://doi.org/10.4171/ifb/471
28. Dong, H.: Gradient estimates for parabolic and elliptic systems from linear laminates. Arch. Ration. Mech. Anal. **205**(1), 119–149 (2012). https://doi.org/10.1007/s00205-012-0501-z
29. Dong, H.: A simple proof of regularity for $C^{1,\alpha}$ interface transmission problems. Ann. Appl. Math. **37**(1), 22–30 (2021)
30. Escauriaza, L.: $W^{2,n}$ a priori estimates for solutions to fully nonlinear equations. Indiana Univ. Math. J. **42**(2), 413–423 (1993). https://doi.org/10.1512/iumj.1993.42.42019
31. Fabes, E.B., Stroock, D.W.: The L^p-integrability of Green's functions and fundamental solutions for elliptic and parabolic equations. Duke Math. J. **51**(4), 997–1016 (1984). https://doi.org/10.1215/S0012-7094-84-05145-7
32. Friedrichs, K.O.: An inequality for potential functions. Am. J. Math. **68**, 581–592 (1946). https://doi.org/10.2307/2371786
33. Friedrichs, K.O.: On the boundary-value problems of the theory of elasticity and Korn's inequality. Ann. Math. **48**(2), 441–471 (1947). https://doi.org/10.2307/1969180
34. Gilbarg, D., Trudinger, N.S.: Elliptic partial differential equations of second order. In: Classics in Mathematics. Springer, Berlin (2001). Reprint of the 1998 edition
35. Huaroto, G., Pimentel, E.A., Rampasso, G.C., Święch, A.: A fully nonlinear degenerate free transmission problem. Ann. PDE **10**(1), Paper No. 5, 30 (2024). https://doi.org/10.1007/s40818-024-00168-x

36. Il'in, V.A., Šišmarev, I.A.: The method of potentials for the problems of Dirichlet and Neumann in the case of equations with discontinuous coefficients. Sibirsk. Mat. Ž. 46–58 (1961)
37. Imbert, C., Silvestre, L.: $C^{1,\alpha}$ regularity of solutions of some degenerate fully non-linear elliptic equations. Adv. Math. **233**, 196–206 (2013). https://doi.org/10.1016/j.aim.2012.07.033
38. Kufner, A., John, O., Fučík, S.: Function spaces. Monographs and Textbooks on Mechanics of Solids and Fluids, Mechanics: Analysis. Noordhoff International Publishing, Leiden, Academia, Prague (1977)
39. Kuusi, T., Mingione, G.: Guide to nonlinear potential estimates. Bull. Math. Sci. **4**(1), 1–82 (2014). https://doi.org/10.1007/s13373-013-0048-9
40. Lee, K.a.: Obstacle problems for the fully nonlinear elliptic operators. ProQuest LLC, Ann Arbor, MI (1998). Thesis (Ph.D.)–New York University
41. Lee, K.A., Shahgholian, H.: Regularity of a free boundary for viscosity solutions of nonlinear elliptic equations. Comm. Pure Appl. Math. **54**(1), 43–56 (2001). https://doi.org/10.1002/1097-0312(200101)54:1<43::AID-CPA2>3.0.CO;2-T
42. Li, Y., Nirenberg, L.: Estimates for elliptic systems from composite material. Comm. Pure Appl. Math. **56**(7), 892–925 (2003). https://doi.org/10.1002/cpa.10079. Dedicated to the memory of Jürgen K. Moser
43. Li, Y.Y., Vogelius, M.: Gradient estimates for solutions to divergence form elliptic equations with discontinuous coefficients. Arch. Ration. Mech. Anal. **153**(2), 91–151 (2000). https://doi.org/10.1007/s002050000082
44. Lieberman, G.M.: The natural generalization of the natural conditions of Ladyzhenskaya and Ural'tseva for elliptic equations. Comm. Part. Differ. Equ. **16**(2-3), 311–361 (1991). https://doi.org/10.1080/03605309108820761
45. Lions, J.L.: Contribution à un problème de M. M. Picone. Ann. Mat. Pura Appl. **41**(4), 201–219 (1956). https://doi.org/10.1007/BF02411668
46. Lions, J.L., Schwartz, L.: Problèmes aux limites sur des espaces fibrés. Acta Math. **94**, 155–159 (1955). https://doi.org/10.1007/BF02392491
47. Maggi, F.: Sets of finite perimeter and geometric variational problems. In: Cambridge Studies in Advanced Mathematics, vol. 135. Cambridge University Press, Cambridge (2012). https://doi.org/10.1017/CBO9781139108133. An introduction to geometric measure theory
48. Oleĭnik, O.A.: Boundary-value problems for linear equations of elliptic parabolic type with discontinuous coefficients. Izv. Akad. Nauk SSSR Ser. Mat. **25**, 3–20 (1961)
49. Picone, M.: Sur un problème nouveau pour l'équation linéaire aux dérivées partielles de la théorie mathématique classique de l'élasticité, pp. 9–11. Colloque sur les équations aux dérivées partielles, CBRM, Bruxelles (1954)
50. Pimentel, E.A., Santos, M.S.: Fully nonlinear free transmission problems. Interfaces Free Bound. **25**(3), 325–342 (2023). https://doi.org/10.4171/ifb/489
51. Pimentel, E.A., Święch, A.: Existence of solutions to a fully nonlinear free transmission problem. J. Differ. Equ. **320**, 49–63 (2022). https://doi.org/10.1016/j.jde.2022.02.057
52. Pimentel, E.A., Teixeira, E.V.: Sharp Hessian integrability estimates for nonlinear elliptic equations: an asymptotic approach. J. Math. Pures Appl. (9) **106**(4), 744–767 (2016). https://doi.org/10.1016/j.matpur.2016.03.010
53. Serrin, J.: Local behavior of solutions of quasi-linear equations. Acta Math. **111**, 247–302 (1964). https://doi.org/10.1007/BF02391014
54. Soria-Carro, M., Stinga, P.R.: Regularity of viscosity solutions to fully nonlinear elliptic transmission problems. Adv. Math. **435**, Paper No. 109353, 52 (2023). https://doi.org/10.1016/j.aim.2023.109353
55. Stampacchia, G.: Su un problema relativo alle equazioni di tipo ellittico del secondo ordine. Ricerche Mat. **5**, 3–24 (1956)
56. Święch, A.: $W^{1,p}$-interior estimates for solutions of fully nonlinear, uniformly elliptic equations. Adv. Differ. Equ. **2**(6), 1005–1027 (1997)
57. Teixeira, E.V.: Universal moduli of continuity for solutions to fully nonlinear elliptic equations. Arch. Ration. Mech. Anal. **211**(3), 911–927 (2014). https://doi.org/10.1007/s00205-013-0688-7

A Concise and Friendly Introduction to the Analysis of Algorithms for Continuous Nonlinear Optimization

Ernesto G. Birgin

Abstract This paper aims to give a brief introduction to the concept of computational complexity in the context of continuous (nonconvex) nonlinear optimization.

1 Introduction

This paper is related to the plenary lecture titled "Complexity results in nonlinear optimization" given by Ernesto G. Birgin at the Brazil-Portugal Joint Meeting on Mathematics (*Encontro Conjunto Brasil-Portugal em Matemática*) that took place from August 14–20, 2022, at the Federal University of Bahia, Salvador, Brazil. In the same way as the presentation, this paper aims to introduce the concept of complexity in continuous (nonconvex) nonlinear optimization to a wide audience of non-specialists and to briefly review some recent results obtained by the author and collaborators.

Optimization is the area of Mathematics that studies the problem of finding $x \in \mathbb{R}^n$ that realizes the smallest possible value of a given objective function, among those points that belong to a predefined feasible region. If the objective function is a continuous function and the feasible region is determined by the points that satisfy a given set of equalities and inequalities defined by continuous functions, then we are dealing with a continuous optimization problem.

Given $f : \mathbb{R}^n \to \mathbb{R}$, $g : \mathbb{R}^n \to \mathbb{R}^p$, $h : \mathbb{R}^n \to \mathbb{R}^m$, and a set $\Omega \subseteq \mathbb{R}^n$, a standard continuous optimization problem can be written as follows

$$\mathop{\text{Minimize}}_{x \in \mathbb{R}^n} f(x) \text{ subject to } h(x) = 0, \; g(x) \leq 0, \; x \in \Omega.$$

If $\Omega = \mathbb{R}^n$ and g and h are absent, the problem is an unconstrained optimization problem. If g and h are absent and $\Omega = \{x \in \mathbb{R}^n \mid \ell \leq x \leq u\}$, where $\ell = (\ell_i)$,

E. G. Birgin (✉)
Department of Computer Science, Institute of Mathematics and Statistics, University of São Paulo, Rua do Matão, Cidade Universitária, São Paulo, SP, Brazil
e-mail: egbirgin@ime.usp.br

M. Abreu et al. (eds.), *Brazil-Portugal Mathematics*, Coimbra Mathematical Texts 4,
https://doi.org/10.1007/978-3-031-98455-6_2

$u = (u_i)$, $\ell_i \in \mathbb{R} \cup \{-\infty\}$, and $u_i \in \mathbb{R} \cup \{+\infty\}$ for $i = 1, \ldots, n$, the problem is a bound-constrained minimization problem. If g is missing the problem is an equality-constrained optimization problem (with or without bound constraints), while if h is missing the problem is an inequality-constrained optimization problem.

Even in the simplest case of unconstrained minimization, a continuous optimization problem can rarely be solved analytically. Therefore, optimization methods are developed. In general, optimization methods are iterative and, given an initial approximation x^0, generate a sequence of iterates $\{x^k\}_{k=0}^{\infty}$ that, hopefully, as k increases, will better approximate a solution or satisfy some desirable condition.

Given an optimization method, there are basically two tasks necessary to show its value. First, it must be shown that it is well defined. That is, that given x^{k-1}, the steps that must be followed at iteration k to define x^k can be carried out. Secondly, the asymptotic properties of the sequence generated by the method must be studied. For example, in unconstrained minimization, a necessary condition for a point to be a local minimizer is that the gradient of the objective function vanishes at the point. Accordingly, a classical asymptotic result consists in showing that $\lim_{k \to +\infty} \nabla f(x^k) = 0$ or that $\lim_{k \in K} \nabla f(x^k) = 0$ for some $K \subset_{\infty} \mathbb{N}$.

Optimization problems are an object of study of Mathematics. However, besides that, they have an enormous number of practical applications in the most diverse areas of science such as Physics, Chemistry, Engineering, Economics, Medicine, and Social Sciences, among others. This practical appeal means that, in practice, optimization methods are equipped with a stopping criterion, since practitioners cannot wait infinite time. When equipped with a stopping criterion, the methods are transformed into algorithms (finite and well-defined sequence of operations). For the example of unconstrained minimization considered in the previous paragraph, given $\epsilon > 0$, a natural stopping criterion would be "to interrupt the execution of the method when finding x^k such that $\|\nabla f(x^k)\| \leq \epsilon$." This is the point where computational complexity comes in to answer the following question: In the worst case, what is the computational cost of finding x^k that satisfies the pre-specified stopping criterion of a given optimization algorithm? It is worth noting that this "computational cost" can be measured in terms of number of arithmetic operations, number of iterations of the algorithm or number of evaluations of the objective function f, among other possibilities. In the last two cases, we are talking about iteration complexity and evaluation complexity, respectively. The field of study known as "computational complexity theory" focuses on the complexity of problems, whereas "analysis of algorithms" is concerned with the computational complexity of algorithms.

The analysis of optimization algorithms for convex and strongly convex problems and for some problems with specific structure, such as fractional linear programming, has a long tradition in the optimization literature. See for example, [21, 25]. However, the convergence theory of optimization methods for nonlinear programming problems consisted basically in the asymptotic study of the infinite sequences generated by the methods until the beginning of the current millennium. In 2006, a paper by Nesterov and Polyak [22], dealing with computational complexity issues in nonconvex continuous optimization, caught the interest of the academic optimization community. Since then, a large number of papers have been published dealing with the computational complexity of different algorithms for a wide variety of nonconvex nonlinear programming problems. In 2022, a book [18] specifically dedicated to this subject was released.

The rest of this paper is organized as follows. In Sect. 2, we present as simply as possible an algorithm for unconstrained minimization and its computational complexity analysis. In Sect. 3, we deal with the computational complexity of an augmented Lagrangian algorithm applied to the solution of the most general continuous optimization problem. The final section presents some perspectives of current and future work.

Notation. The symbol $\|\cdot\|$ denotes the Euclidean norm of vectors and matrices. For $v \in \mathbb{R}^n$, $v_+ = (\max\{0, v_1\}, \dots, \max\{0, v_n\})^T$. If $\ell, u \in \mathbb{R}^n$, $[\ell, u]$ denotes the box defined by $\{x \in \mathbb{R}^n \mid \ell \le x \le u\}$, while $P_{[\ell,u]}(x)$ denotes the projection operator of an arbitrary point $x \in \mathbb{R}^n$ onto $[\ell, u]$. If $K = \{k_1, k_2, \dots\} \subseteq \mathbb{N}$ (with $k_j < k_{j+1}$ for all j), we denote $K \subset_\infty \mathbb{N}$. Given $\phi : \mathbb{R}^n \to \mathbb{R}^m$, $\phi = (\phi_1, \dots, \phi_m)^T$, we denote $\nabla\phi(x) = (\nabla\phi_1(x), \dots, \nabla\phi_m(x))$. $\mathbb{R}^n_+ = \{x \in \mathbb{R}^n \mid x \ge 0\}$. Given two functions $a : \mathbb{R} \to \mathbb{R}$ and $b : \mathbb{R} \to \mathbb{R}$, we say $a(\epsilon) = O(b(\epsilon))$ as $\epsilon \to 0$ if there exist positive numbers δ and M such that $|a(\epsilon)| \le Mb(x)$ for all $0 < \epsilon < \delta$.

2 A Didactic Example of Computational Complexity for Unconstrained Minimization

In this section we consider an unconstrained minimization problem given by

$$\underset{x \in \mathbb{R}^n}{\text{Minimize}}\ f(x), \tag{1}$$

where $f : \mathbb{R}^n \to \mathbb{R}$. To tackle problem (1), we introduce a method based on "cubic regularization" that, given an initial guess $x^0 \in \mathbb{R}^n$, generates a sequence $\{x^k\}_{k=0}^{\infty}$. We show that, given a tolerance $\epsilon > 0$, the method uses a finite number of iterations of order $\epsilon^{-3/2}$ to find a point x^k such that $\|\nabla f(x^k)\| \le \epsilon$. In fact, we show more than that. We show that the (finite) number of iterations such that $\|\nabla f(x^k)\| > \epsilon$ is of the order of $\epsilon^{-3/2}$. The results are different because from the former one it would follow that $\lim_{k\in K} \nabla f(x^k) = 0$ for some $K \subset_\infty \mathbb{N}$, while from the latter one it follows that $\lim_{k\to+\infty} \nabla f(x^k) = 0$. The complexity results on the number of iterations also follow for the number of evaluations of f, because in the introduced method the number of evaluations of f per iterations is $O(1)$ with respect to ϵ.

The method follows below.

Method 1 Let $\alpha > 0$ and $x^0 \in \mathbb{R}^n$ be given. Set $k \leftarrow 1$.

Step 1. Define $x^k = x^{k-1} + s^k$, where s^k is such that

$$f(x^k) \le f(x^{k-1}) - \alpha\|s^k\|^3. \tag{2}$$

Step 2. Set $k \leftarrow k + 1$ and go to Step 1.

At this point, Method 1 is not very elucidative, because we did not say yet in what way an s^k satisfying (2) can be computed. (In fact we did not even mention under what conditions an s^k satisfying (2) exists, so it is not clear yet under what hypotheses the method is well defined.) However, it is interesting to understand what properties Method 1 has under suitable assumptions.

Let us assume that there exists $\gamma > 0$ such that

$$\sqrt{\frac{\|\nabla f(x^k)\|}{\gamma}} \le \|s^k\| \tag{3}$$

for all $k \in \mathbb{N}$. (Later in this section we will give sufficient conditions for this assumption to be satisfied.) From (2) and (3), it trivially follows that, for all $k \in \mathbb{N}$,

$$f(x^k) \le f(x^{k-1}) - c\|\nabla f(x^k)\|^{3/2} \tag{4}$$

with $c = \alpha/\gamma^{3/2}$.

Assume now that, given a tolerance $\epsilon > 0$, we desire to stop Method 1 the first time an iterate x^k satisfying $\|\nabla f(x^k)\| \le \epsilon$ is computed. Assume the method is at (the end of) iteration k_{current} and it did not stop. It means that $\|\nabla f(x^k)\| > \epsilon$ for $k = 1, \dots, k_{\text{current}}$. Thus, by (4), $f(x^k) \le f(x^{k-1}) - c\epsilon^{3/2}$ for $k = 1, \dots, k_{\text{current}}$, i.e. the value of the objective function decreased at least $c\epsilon^{3/2}$ at every iteration already executed. At this point there are two alternatives. Or the method is in the way to compute a sequence $\{x^k\}_{k=0}^{\infty}$ such that $\lim_{k \in K} f(x^k) = -\infty$ for some $K \subset_{\infty} \mathbb{N}$ (in which case f is unbounded below) or there exists $f_{\text{low}} \in \mathbb{R}$ such that $f(x^k) \ge f_{\text{low}}$ for all $k \in \mathbb{N}$ and

$$k_{\text{current}} \le k_\epsilon := \left\lfloor \frac{f(x^0) - f_{\text{low}}}{c\epsilon^{3/2}} \right\rfloor = O(\epsilon^{-3/2}).$$

We name the assumption "there exists $f_{\text{low}} \in \mathbb{R}$ such that $f(x^k) \ge f_{\text{low}}$ for all $k \in \mathbb{N}$" **Assumption A1** from now on. This assumption is an assumption in the sequence generated by the method, which is undesired. On the other hand, it holds trivially if there exists f_{low} such that $f(x) \ge f_{\text{low}}$ for all $x \in \mathbb{R}^n$.

Up to this point, we have shown that, under assumption (3) and Assumption A1, given $\epsilon > 0$, there exists $k \le k_\epsilon + 1$ such that $\|\nabla f(x^k)\| \le \epsilon$ and that the number of iterations such that $\|\nabla f(x^k)\| > \epsilon$ is limited by k_ϵ. From the former it follows $\lim_{k \in K} \nabla f(x^k) = 0$ for some $K \subset_{\infty} \mathbb{N}$, while from the latter it follows $\lim_{k \to +\infty} \nabla f(x^k) = 0$. It remains to show how to compute s^k at each iteration k using $O(1)$ functional evaluations. By showing this we will also address the satisfaction of (3) and we will reveal from where the qualifier "cubic regularized" for Method 1 comes from.

There are several alternatives to compute, at iteration k, a step s^k satisfying the sufficient descent condition (2) plus (3). A simple choice will be present here for didactical purposes. Consider the cubic regularized second-order Taylor polynomial of f around x^{k-1} given by

$$M_k(s,\sigma) := T_k(s) + \frac{\sigma}{3}\|s\|^3,$$

where

$$T_k(s) := f(x^{k-1}) + \nabla f(x^{k-1})^T s + \frac{1}{2} s^T \nabla^2 f(x^{k-1}) s$$

and $\sigma > 0$ plays the role of a regularization parameter. The procedure to compute s^k is described by the method below.

Method 2 Let $\sigma_{k,1} \geq \sigma_{\text{low}} > 0$ and $\theta > 0$ be given (with σ_{low} and θ being the same for all k.) Let $\ell \leftarrow 1$.

Step 1. Compute $s^{k,\ell}$ such that

$$M_k(s^{k,\ell}, \sigma_{k,\ell}) \leq M_k(0, \sigma_{k,\ell}) \tag{5}$$

and

$$\|\nabla_s M_k(s^{k,\ell}, \sigma_{k,\ell})\| \leq \theta \|s^{k,\ell}\|^2. \tag{6}$$

Step 2. If (2) does not hold with $s^k \equiv s^{k,\ell}$ then define $\sigma_{k,\ell+1} := 2\sigma_{k,\ell}$, set $\ell \leftarrow \ell + 1$ and go to Step 1.

Step 3. Define $\sigma_k := \sigma_{k,\ell}$ and $s^k := s^{k,\ell}$.

The first task is to show that Method 2 is well defined and that it is in fact an algorithm, i.e. that it stops in a finite number of iterations (having computed a step s^k that satisfies (2) plus (5, 6) as desired.) The fact that the method is well defined comes from the fact that $s^{k,\ell}$ satisfying (5, 6) can be computed for all k and ℓ. This is because the models $M_k(s, \sigma_{k,\ell})$ with $\sigma_{k,\ell} > 0$ have bounded level sets and, thus, have at least one global minimizer. At the global minimizer, the functional value is upper bounded by $M_k(0, \sigma_{k,\ell})$ and the gradient vanishes. Therefore, (5, 6) hold. Moreover, for given k and ℓ, the step $s^{k,\ell}$ can be computed in finite time by a monotone method for unconstrained minimization with the property that the accumulation points of the sequence generated by the method are first-order stationary, starting from $s = 0$. The combination of the method monotonicity and the initial guess choice guarantees the satisfaction of (5). Concerning (6), if $s = 0$ is stationary of minimizing $M_k(s, \sigma_{k,\ell})$, then the method stops at the initial point and (6) holds. Let us say then that $s = 0$ is not stationary. Since model $M_k(s, \sigma_{k,\ell})$ with $\sigma_{k,\ell} > 0$ has bounded level sets, then the sequence $\{s^{k,\ell,j}\}_{j=0}^{\infty}$ generated by the method has a stationary accumulation point z, which is not zero. Then, there exists $J \subset_\infty \mathbb{N}$ such that, for $j \in J$ sufficiently large, $\|\nabla_s M_k(s^{k,\ell,j}, \sigma_{k,\ell})\| \leq \frac{1}{2}\theta\|z\|^2$ and $\frac{1}{2}\theta\|z\|^2 \leq \theta\|s^{k,\ell,j}\|^2$, i.e. $s^{k,\ell,j}$ satisfies (6). It is worth noting that the task of computing $s^{k,\ell}$ does not depend on ϵ and does not require evaluations of the objective function f. Thus, the entire set of operations carried out by Method 2 has no effect on the complexity of Method 1.

The task of showing that Method 2 stops in a finite number of iterations requires to assume that there exist non-negative constants ξ_1 and ξ_2 such that the second-order Taylor polynomials T_k satisfy

$$f(x^{k-1} + s^{k,\ell}) - T_k(s^{k,\ell}) \le \xi_1 \|s^{k,\ell}\|^3 \tag{7}$$

and

$$\|\nabla f(x^{k-1} + s^{k,\ell}) - \nabla_s T_k(s^{k,\ell})\| \le \xi_2 \|s^{k,\ell}\|^2 \tag{8}$$

for all k and ℓ. We name it **Assumption A2**. This assumption is fulfilled when f is three times continuously differentiable on $\mathbb{R}^n$ and the third-order derivative of f is bounded or when f is twice continuously differentiable and the second-order derivative is Lipschitz continuous; see, for example, [4]. With these assumptions, in particular using (7), (5), and the fact that $M_k(0, \cdot) = f(x^{k-1})$, we have that, if

$$\sigma_{k,\ell} \ge 3(\xi_1 + \alpha), \tag{9}$$

then

$$\begin{aligned} f(x^{k-1} + s^{k,\ell}) &\le T_k(s^{k,\ell}) + \xi_1 \|s^{k,\ell}\|^3 \\ &= T_k(s^{k,\ell}) + \tfrac{\sigma_{k,\ell}}{3} \|s^{k,\ell}\|^3 - (\tfrac{\sigma_{k,\ell}}{3} - \xi_1)\|s^{k,\ell}\|^3 \\ &= M_k(s^{k,\ell}, \sigma_{k,\ell}) - (\tfrac{\sigma_{k,\ell}}{3} - \xi_1)\|s^{k,\ell}\|^3 \\ &\le M_k(0, \sigma_{k,\ell}) - \alpha\|s^{k,\ell}\|^3 \\ &= f(x^{k-1}) - \alpha\|s^{k,\ell}\|^3, \end{aligned}$$

i.e. that $s^{k,\ell}$ satisfies (2). Since $\sigma_{k,1} \ge \sigma_{\text{low}}$ and $\sigma_{k,\ell} = 2\sigma_{k,\ell-1}$ for $\ell = 2, 3, \ldots$ then it is clear that Method 2 achieves a sufficiently large value for the regularization parameter in at most $\lceil \log_2((3(\xi_1 + \alpha))/\sigma_{\text{low}}) + 1 \rceil$ iterations, a quantity that does not depend on ϵ. At each iteration ℓ, Method 2 needs to check if the computed trial step $s^{k,\ell}$ satisfies (2). This verification involves an evaluation of f.

Now observe that, by the increasing rule of the regularization parameter in Method 2 (i.e. doubling it) and the fact that $\sigma_{k,\ell} \ge 3(\xi_1 + \alpha)$ makes $s^{k,\ell}$ to satisfy the sufficient descent condition (2), it holds that

$$\sigma_k < \sigma_{\max} := 6(\xi_1 + \alpha) \tag{10}$$

for all k. We now aim to show that (10), (8), and (6) imply (3). Let us write

$$\|\nabla f(x^k)\| = \|\nabla f(x^{k-1} + s^k)\| = \|\nabla f(x^{k-1} + s^k) - \nabla_s M_k(s^k, \sigma_k) + \nabla_s M_k(s^k, \sigma_k)\|. \tag{11}$$

Now, since

$$\nabla_s M_k(s, \sigma) = \nabla_s T_k(s) + \sigma \|s\|^2 \frac{s}{\|s\|}, \tag{12}$$

by substituting (12) in (11), applying the triangle inequality, and then using (8, 10, 6), we have

$$\begin{aligned}
\|\nabla f(x^k)\| &= \|\nabla f(x^{k-1} + s^k) - \nabla_s T_k(s^k) - \sigma_k \|s^k\|^2 \tfrac{s^k}{\|s^k\|} + \nabla_s M_k(s^k, \sigma_k)\| \\
&\le \|\nabla f(x^{k-1} + s^k) - \nabla_s T_k(s^k)\| + \|\sigma_k \|s^k\|^2\| + \|\nabla_s M_k(s^k, \sigma_k)\| \\
&\le (\xi_2 + \sigma_{\max} + \theta)\|s^k\|^2.
\end{aligned}$$

So (3) holds with $\gamma = \xi_2 + \sigma_{\max} + \theta$.

Summarizing, we showed in this section an algorithm that, given $f : \mathbb{R}^n \to \mathbb{R}$ and $\epsilon > 0$, finds a point x^k that satisfies $\|f(x^k)\| \le \epsilon$ in a finite number of iterations of order $\epsilon^{-3/2}$. For that, Assumptions A1 and A2 are required, and sufficient conditions for their satisfaction are that (i) f is bounded from below by f_{low} in $\mathbb{R}^n$ and that (ii)a f is three times continuously differentiable on $\mathbb{R}^n$ and the third-order derivative of f is bounded or (ii)b f is twice continuously differentiable and the second-order derivative is Lipschitz continuous.

It is important to mention that the derivations presented in this section are nowadays standard in the literature and can be found, among many others, in, for example, [1, 5–8, 10–12, 14–17, 22] applied in proofs of complexity results of algorithms for unconstrained minimization, minimization with bound-constraints or convex sets, and nonlinear programming in general. Moreover, the presented results can also be extended, as in fact they already were, for arbitrary norms, models of order p (other than Taylor polynomials) with regularization of order $p + 1$ and to find stationary points of order q instead of first-order stationary points only.

3 The Case of Augmented Lagrangians for Nonlinear Optimization

In this section we consider the constrained problem given by

$$\underset{x \in \mathbb{R}^n}{\text{Minimize}}\ f(x) \text{ subject to } h(x) = 0,\ g(x) \le 0,\ \ell \le x \le u \tag{13}$$

where $f : \mathbb{R}^n \to \mathbb{R}$, $g : \mathbb{R}^n \to \mathbb{R}^p$ and $h : \mathbb{R}^n \to \mathbb{R}^m$ are continuously differentiable and $\ell, u \in \mathbb{R}^n$.

Augmented Lagrangians represent a well-established family of methods for solving nonlinear programming problems of the form (13). The computational complexity of different augmented Lagrangian methods was studied in [13, 19, 26]. In [13], the complexity of Algencan was analyzed. Algencan [2, 3, 9] is an implementation of a safeguarded augmented Lagrangian method relying on the Powell-Hestenes-Rockafellar augmented Lagrangian function [20, 23, 24]. Each iteration of Algencan consists of (a) minimization of the objective function plus a term that penalizes violation of the constraints with respect to appropriate shifted tolerances and (b) updating of the Lagrange multipliers approximations and the penalty parameter.

The Powell-Hestenes-Rockafellar augmented Lagrangian function is given by

$$L_\rho(x, \lambda, \mu) := f(x) + \frac{\rho}{2}\left[\sum_{i=1}^{m}\left(h_i(x) + \frac{\lambda_i}{\rho}\right)^2 + \sum_{i=1}^{p}\left(g_i(x) + \frac{\mu_i}{\rho}\right)_+^2\right]$$

for all $x \in [\ell, u]$, $\rho > 0$, $\lambda \in \mathbb{R}^m$, and $\mu \in \mathbb{R}^p_+$, where ρ is the penalty parameter and λ and μ represent the Lagrange multipliers associated with the equality constraints $h(x) = 0$ and the inequality constraints $g(x) \le 0$, respectively.

Method 3 below describes Algencan.

Method 3 Assume that $x^0 \in \mathbb{R}^n$, $\lambda_{\min} < \lambda_{\max}$, $\bar{\lambda}^1 \in [\lambda_{\min}, \lambda_{\max}]^m$, $\mu_{\max} > 0$, $\bar{\mu}^1 \in [0, \mu_{\max}]^p$, $\rho_1 > 0$, $\gamma > 1$, $0 < \tau < 1$, and $\{\epsilon_k\}_{k=1}^{\infty}$ are given. Initialize $k \leftarrow 1$.

Step 1. Compute $x^k \in [\ell, u]$ satisfying

$$\left\| P_{[\ell,u]}\left(x^k - \nabla L_{\rho_k}(x^k, \bar{\lambda}^k, \bar{\mu}^k)\right) - x^k \right\|_\infty \le \epsilon_k \tag{14}$$

by approximately solving

$$\underset{x \in \mathbb{R}^n}{\text{Minimize}}\ L_{\rho_k}(x, \bar{\lambda}^k, \bar{\mu}^k) \text{ subject to } x \in [\ell, u].$$

Step 2. Define

$$V^k = \min\left\{-g(x^k), \frac{\bar{\mu}^k}{\rho_k}\right\}.$$

If $k = 1$ or

$$\max\left\{\|h(x^k)\|_\infty, \|V^k\|_\infty\right\} \le \tau \max\left\{\|h(x^{k-1})\|_\infty, \|V^{k-1}\|_\infty\right\},$$

choose $\rho_{k+1} \ge \rho_k$. Otherwise, choose $\rho_{k+1} \ge \gamma\rho_k$.

Step 3. Compute

$$\lambda^{k+1} = \bar{\lambda}^k + \rho_k h(x^k) \text{ and } \mu^{k+1} = \left(\bar{\mu}^k + \rho_k g(x^k)\right)_+.$$

Compute $\bar{\lambda}^{k+1} \in [\lambda_{\min}, \lambda_{\max}]^m$ and $\bar{\mu}_i^{k+1} \in [0, \mu_{\max}]^p$. Set $k \leftarrow k + 1$ and go to Step 1.

Algencan was introduced in [2, 3] and it is fully described in the book [9]. In particular, [9] describes the asymptotic convergence theory of Algencan. On the other hand, [13] complements Algencan's theory presented in [9] by presenting its worst-case iteration and evaluation complexity analysis. In this section we summarize the Algencan worst-case complexity analysis presented in [13].

Method 3 has interesting properties for the case of sequences of arbitrary tolerances $\{\epsilon_k\}_{k=1}^{\infty}$, which are used as tolerance for the approximate solution of the subproblems in Step 1; see [9] for details. However, in the present work, we are interested in the case $\epsilon_k \to 0$. For that reason, from here on, we will analyze this case only. When Method 3 is applied to problem (13), the method generates sequences of primal iterands $\{x^k\}$ and sequences of Lagrange multipliers $\{\lambda^k\}$ and $\{\mu^k\}$. If $\lim_{k\in K} x^k = x^*$ for some $K \subset_\infty \mathbb{N}$, then there are two possibilities for x^*. Either x^* is *infeasible* and satisfies the first-order optimality conditions for the problem of minimizing the squared infeasibility given by

$$\underset{x\in[\ell,u]}{\text{Minimize}} \; \|h(x)\|^2 + \|g(x)_+\|^2$$

or x^* is *feasible* and satisfies the sequential optimality condition AKKT of problem (13) given by

$$\lim_{k\in K} \left\| P_{[\ell,u]}\left(x^k - \left(\nabla f(x^k) + \nabla h(x^k)\lambda^{k+1} + \nabla g(x^k)\mu^{k+1}\right)\right) - x^k \right\| = 0$$

and

$$\lim_{k\in K} \max\left\{\|h(x^k)\|, \|\min\{-g(x^k), \mu^{k+1}\}\|\right\} = 0.$$

As in this work we are interested in complexity results associated with Method 3, our objective is to transform Method 3 into an algorithm incorporating stopping criteria associated with the two possibilities mentioned in the previous paragraph and to quantify the computational cost that the resulting algorithm consumes to stop due to either of the two criteria.

Stating the complexity results requires a couple of lemmas that we present below. Proofs of the lemmas can be found at [13]. We record here that these lemmas involve hypotheses on problem (13), namely, the bondedness of $[\ell, u]$ and the continuity of f, g h, and their gradients. The main complexity results of Method 3 are stated in the sequel. Additional results can be found in [13].

Lemma 4 ([13, Lemma 3.2]) *There exists $c_{\rm big} > 0$ such that, for all $k \geq 1$,*

$$\max\{\|h(x^k)\|_\infty, \|V_k\|_\infty\} \leq c_{\rm big}.$$

Lemma 5 ([13, Lemma 3.3]) *There exist $c_{\rm lips} > 0$ and $c_f > 0$ such that, for all $x \in [\ell, u]$, $\lambda \in [\lambda_{\min}, \lambda_{\max}]^m$, and $\mu \in [0, \mu_{\max}]^p$, one has*

$$\|\nabla h(x)\| \|\lambda\| + \|\nabla g(x)\| \|\mu\| \leq c_{\rm lips}$$

and

$$\|\nabla f(x)\| \leq c_f.$$

The theorem that follows presents a complexity result that limits the number of iterations needed to compute a triple $(x^k, \lambda^{k+1}, \mu^{k+1})$ that either satisfies (16, 17) or (18, 19, 20). The conditions (16, 17) relate to an infeasible point x^k that is stationary of the sum of the squared infeasibilities subject to the box constraints. The relation $\delta > \delta_{\rm low}$ suggests that x^k is "very infeasible" and "nearly stationary". The conditions (18, 19, 20) correspond to the approximate satisfaction of the KKT conditions by the triple $(x^k, \lambda^{k+1}, \mu^{k+1})$ with tolerance ϵ for optimality and tolerance δ for feasibility and complementarity. In the bound on the number of iterations, the constant $N(\delta_{\rm low}, \epsilon)$ appears such that $\epsilon_k \leq \min\{\epsilon, \delta_{\rm low}\}/4$ for all $k \geq N(\delta_{\rm low}, \epsilon)$. As we mentioned above, $\epsilon_k \to 0$. It would be acceptable to assume $\epsilon_k \leq \min\{\epsilon, \delta_{\rm low}\}/4$ for all k, in which case $N(\delta_{\rm low}, \epsilon) = 1$. However, from a practical point of view, it is important to allow successive subproblems of the augmented Lagrangian method to be solved approximately, with varying tolerance ϵ_k. The complexity result shows how the bound on the number of iterations is affected by the choice of these tolerances.

Theorem 6 ([13, Theorem. 3.5]) *Let $\delta > 0$, $\delta_{\rm low} \in (0, \delta)$, and $\epsilon > 0$ be given. Let $N(\delta_{\rm low}, \epsilon)$ be such that $\epsilon_k \leq \min\{\epsilon, \delta_{\rm low}\}/4$ for all $k \geq N(\delta_{\rm low}, \epsilon)$. Then, after at most*

$$\max\left\{N(\delta_{\rm low}, \epsilon), \left\lceil \frac{\log(\delta/c_{\rm big})}{\log(\tau)} \right\rceil \times \left\lceil \frac{\log\left(\rho_{\max}/\rho_1\right)}{\log(\gamma)} \right\rceil\right\}$$

iterations, where

$$\rho_{\max} = \max\left\{1, \frac{4c_{\rm lips}}{\delta_{\rm low}}, \frac{\mu_{\max}}{\delta}, \frac{4c_f}{\delta_{\rm low}}\right\}, \tag{15}$$

we obtain an iteration k such that one of the following two facts takes place:

1. *The iterate $x^k \in [\ell, u]$ verifies*

$$\left\| P_{[\ell,u]}\left(x^k - \nabla\left[\|h(x^k)\|^2 + \|g(x^k)_+\|^2\right]\right) - x^k \right\|_\infty \leq \delta_{\rm low} \tag{16}$$

and

$$\max\{\|h(x^k)\|_\infty, \|g(x^k)_+\|_\infty\} > \delta. \tag{17}$$

2. *The multipliers $\lambda^{k+1} \in \mathbb{R}^m$ and $\mu^{k+1} \in \mathbb{R}^p_+$ are such that*

$$\left\| P_{[\ell,u]}\left(x^k - \left(\nabla f(x^k) + \nabla h(x^k)\lambda^{k+1} + \nabla g(x^k)\mu^{k+1}\right)\right) - x^k \right\|_\infty \leq \epsilon, \tag{18}$$

$$\|h(x^k)\|_\infty \leq \delta, \ \|g(x^k)_+\|_\infty \leq \delta, \tag{19}$$

and, for all $j = 1, \ldots, p$,

$$\mu_j^{k+1} = 0 \text{ whenever } g_j(x^k) < -\delta. \tag{20}$$

The previous theorem limits the number of iterations of the algorithm, but the result is incomplete if the cost of solving the subproblem at each iteration is not mentioned. The minimized augmented Lagrangian function at each iteration depends on the penalty parameter ρ_k. In the next theorem we assume that the cost of solving each subproblem is known and depends on ρ_k. The theorem presents a bound for the total number of iterations of the algorithm used to solve the subproblems and the total number of evaluations of the functions f, g, and h that define the original problem.

Theorem 7 ([13, Theorem 3.6]) *In addition to the hypotheses of Theorem 6, assume that there exist $\bar{c}_{\text{inner}} > 0$, $v > 0$, and $q > 0$, where $\bar{c}_{\text{inner}}$ only depends on $\lambda_{\min}$, $\lambda_{\max}$, $\mu_{\max}$, ℓ, u, and characteristics of the functions f, h, and g, such that the number of inner iterations, function and derivative evaluations that are necessary to obtain (14) is bounded above by $\bar{c}_{\text{inner}}\, \rho_k^v \epsilon_k^{-q}$. Then, the number of inner iterations, function evaluations, and derivative evaluations that are necessary to obtain k such that (16) and (17) hold or (18), (19) and (20) hold is bounded above by*

$$\bar{c}_{\text{inner}}\, \rho_{\max}^v \epsilon_{\min,3}^{-q} \max\left\{N(\delta_{\text{low}}, \epsilon), \left\lceil \frac{\log(\delta/c_{\text{big}})}{\log(\tau)}) \right\rceil \times \left\lceil \frac{\log(\rho_{\max}/\rho_1)}{\log(\gamma)} \right\rceil \right\},$$

where $\rho_{\max}$ is given by (15) and

$$\epsilon_{\min,3} = \min\left\{\epsilon_k \mid k \le \max\left\{N(\delta_{\text{low}}, \epsilon), \left\lceil \frac{\log(\delta/c_{\text{big}})}{\log(\tau)}) \right\rceil \times \left\lceil \frac{\log(\rho_{\max}/\rho_1)}{\log(\gamma)} \right\rceil \right\}\right\}.$$

The analysis of the algorithm that Algencan uses to solve the subproblems, meeting the hypotheses of Theorem 7, is presented in [13].

It is worth keeping in mind that the hypotheses used on problem (13) to obtain the complexity results mentioned above are minimal and they do not include any constraint qualification. Consider an algorithm that, given a problem of the form (13) and a tolerance $\epsilon > 0$, checks whether $(x, \lambda, \mu) = 0$ satisfies first-order optimality conditions with precision ϵ. If it satisfies, it returns $(x, \lambda, \mu) = 0$. Otherwise, it declares that it failed. With the hypothesis "$(x, \lambda, \mu) = 0$ is a first-order stationary point of the problem", we would prove that the method finds a solution to problem (13) with computational complexity $O(1)$. It would definitely be the most efficient method in the world, but it would be of little use.

4 Conclusions and Perspectives for Future Work

In this text, we introduced the concept of computational complexity or analysis of algorithms in the area of continuous nonconvex nonlinear optimization. We illustrated the idea with the simplest possible case of unconstrained minimization and concluded by addressing the more general case of an augmented Lagrangian algorithm for nonlinear programming. In between, many problems and algorithms can and have been considered in the literature. A careful look at the references in this work or a search on the author's web page, his Google Scholar profile or the references of the book [18] would be a good starting point for the interested reader.

One line of research widely used in the literature consists of, for a given problem, trying to develop the algorithm with the best possible complexity. Two important points must be considered when this point of view is used. The first is related to the hypotheses that the algorithm (with potentially low complexity) needs to find points with the desired properties. In other words, doesn't the search for a competitive complexity leave out many problems that do not satisfy the necessary hypotheses? (See [7, Table 2].) The second question concerns the applicability of the proposed method. Often, an algorithm with low worst-case complexity is known to be useless in practice, because it suffers from known issues, such as generating too short steps far from a solution.

The effective contribution of complexity analysis to the development of novel algorithms for (nonconvex) nonlinear programming problems that are clean, easy to understand and implement, and have significantly better performance than existing, well-established ones remains to be verified. For the reasons mentioned in the previous paragraph, this author believes that the genuine and already verified contribution of complexity analysis of continuous optimization algorithms lies in deepening or closing a gap in the study of existing consolidated algorithms. In-depth knowledge of existing algorithms helps to use them in the best possible way, to better interpret their behavior and, potentially, to incorporate improvements.

References

1. Amaral, V.S., Andreani, R., Birgin, E.G., Marcondes, D.S., Martínez, J.M.: On complexity and convergence of high-order coordinate descent algorithms for smooth nonconvex box-constrained minimization. J. Global Optim. **84**(3), 527–561 (2022). https://doi.org/10.1007/s10898-022-01168-6
2. Andreani, R., Birgin, E.G., Martínez, J.M., Schuverdt, M.L.: On augmented Lagrangian methods with general lower-level constraints. SIAM J. Optim. **18**(4), 1286–1309 (2007). https://doi.org/10.1137/060654797
3. Andreani, R., Birgin, E.G., Martínez, J.M., Schuverdt, M.L.: Augmented Lagrangian methods under the constant positive linear dependence constraint qualification. Math. Program. **111**(1–2), 5–32 (2008). https://doi.org/10.1007/s10107-006-0077-1
4. Bertsekas, D.P.: Nonlinear Programming. Athena Scientific Optimization and Computation Series, 2nd edn. Athena Scientific, Belmont, MA (1999)
5. Birgin, E.G., Bueno, L.F., Martínez, J.M.: On the complexity of solving feasibility problems with regularized models. Optim. Methods Softw. **37**(2), 405–424 (2022). https://doi.org/10.1080/10556788.2020.1786564
6. Birgin, E.G., Gardenghi, J.L., Martínez, J.M., Santos, S.A.: On the use of third-order models with fourth-order regularization for unconstrained optimization. Optim. Lett. **14**(4), 815–838 (2020). https://doi.org/10.1007/s11590-019-01395-z
7. Birgin, E.G., Gardenghi, J.L., Martínez, J.M., Santos, S.A., Toint, P.L.: Evaluation complexity for nonlinear constrained optimization using unscaled KKT conditions and high-order models. SIAM J. Optim. **26**(2), 951–967 (2016). https://doi.org/10.1137/15M1031631

8. Birgin, E.G., Gardenghi, J.L., Martínez, J.M., Santos, S.A., Toint, P.L.: Worst-case evaluation complexity for unconstrained nonlinear optimization using high-order regularized models. Math. Program. **163**(1–2), 359–368 (2017). https://doi.org/10.1007/s10107-016-1065-8
9. Birgin, E.G., Martínez, J.M.: Practical augmented Lagrangian methods for constrained optimization. In: Fundamentals of Algorithms, vol. 10. Society for Industrial and Applied Mathematics (SIAM), Philadelphia, PA (2014). https://doi.org/10.1137/1.9781611973365
10. Birgin, E.G., Martínez, J.M.: The use of quadratic regularization with a cubic descent condition for unconstrained optimization. SIAM J. Optim. **27**(2), 1049–1074 (2017). https://doi.org/10.1137/16M110280X
11. Birgin, E.G., Martínez, J.M.: On regularization and active-set methods with complexity for constrained optimization. SIAM J. Optim. **28**(2), 1367–1395 (2018). https://doi.org/10.1137/17M1127107
12. Birgin, E.G., Martínez, J.M.: A Newton-like method with mixed factorizations and cubic regularization for unconstrained minimization. Comput. Optim. Appl. **73**(3), 707–753 (2019). https://doi.org/10.1007/s10589-019-00089-7
13. Birgin, E.G., Martínez, J.M.: Complexity and performance of an augmented Lagrangian algorithm. Optim. Methods Softw. **35**(5), 885–920 (2020). https://doi.org/10.1080/10556788.2020.1746962
14. Birgin, E.G., Martínez, J.M.: Block coordinate descent for smooth nonconvex constrained minimization. Comput. Optim. Appl. **83**(1), 1–27 (2022). https://doi.org/10.1007/s10589-022-00389-5
15. Cartis, C., Gould, N.I.M., Toint, P.L.: On the complexity of steepest descent, Newton's and regularized Newton's methods for nonconvex unconstrained optimization problems. SIAM J. Optim. **20**(6), 2833–2852 (2010). https://doi.org/10.1137/090774100
16. Cartis, C., Gould, N.I.M., Toint, P.L.: Adaptive cubic regularisation methods for unconstrained optimization. Part II: worst-case function- and derivative-evaluation complexity. Math. Program. **130**(2), 295–319 (2011). https://doi.org/10.1007/s10107-009-0337-y
17. Cartis, C., Gould, N.I.M., Toint, P.L.: An adaptive cubic regularization algorithm for nonconvex optimization with convex constraints and its function-evaluation complexity. IMA J. Numer. Anal. **32**(4), 1662–1695 (2012). https://doi.org/10.1093/imanum/drr035
18. Cartis, C., Gould, N.I.M., Toint, P.L.: Evaluation Complexity of Algorithms for Nonconvex Optimization: Theory, Computation, and Perspectives. In: MOS-SIAM Series on Optimization, vol. 30. Society for Industrial and Applied Mathematics (SIAM), Philadelphia, PA; Mathematical Optimization Society, Philadelphia, PA (2022). https://doi.org/10.1137/1.9781611976991
19. Grapiglia, G.N., Yuan, Y.X.: On the complexity of an augmented Lagrangian method for nonconvex optimization. IMA J. Numer. Anal. **41**(2), 1546–1568 (2021). https://doi.org/10.1093/imanum/draa021
20. Hestenes, M.R.: Multiplier and gradient methods. J. Optim. Theory Appl. **4**, 303–320 (1969). https://doi.org/10.1007/BF00927673
21. Nemirovsky, A.S., Yudin, D.B.: Problem complexity and method efficiency in optimization. Wiley-Interscience Series in Discrete Mathematics. Wiley, New York, Translated from the Russian and with a preface by E. A Wiley-Interscience Publication, R. Dawson (1983)
22. Nesterov, Y., Polyak, B.T.: Cubic regularization of Newton method and its global performance. Math. Program. **108**(1), 177–205 (2006). https://doi.org/10.1007/s10107-006-0706-8
23. Powell, M.J.D.: A method for nonlinear constraints in minimization problems. In: Optimization (Symposium, University, Keele, Keele, 1968), pp. 283–298. Academic Press, London-New York (1969)
24. Rockafellar, R.T.: Augmented Lagrange multiplier functions and duality in nonconvex programming. SIAM J. Control **12**, 268–285 (1974)
25. Vavasis, S.A.: Nonlinear Optimization: Complexity Issues. International Series of Monographs on Computer Science, vol. 8. The Clarendon Press, Oxford University Press, New York (1991)
26. Xie, Y., Wright, S.J.: Complexity of Proximal Augmented Lagrangian for Nonconvex Optimization with Nonlinear Equality Constraints. J. Sci. Comput. **86**(3), Article No. 38, 30 (2021). https://doi.org/10.1007/s10915-021-01409-y

Remarks on Non-formal Deformation Quantization of Poisson Manifolds

Alejandro Cabrera and Rui Loja Fernandes

Abstract We survey the problem of non-formal deformation quantization of Poisson manifolds. This survey is a written version of the talk given by the second author at the Bahia meeting and includes ongoing, unpublished work, by the authors.

1 Introduction

Heuristically, the term *quantization* refers to a variety of procedures in which one aims to associate to a classical system the data underlying a quantum system. In particular, to classical observables described by functions on a Poisson manifold (see Sect. 2) one aims to associate quantum observables consisting of operators on a Hilbert space. This correspondence, denoted

$$f \mapsto Q_\hbar(f),$$

depends on Planck's constant $\hbar$, which maybe interpreted as a scale or a deformation parameter. The notion of *deformation quantization* arises from considering a *star product* structure $\star_\hbar$ defined by the relation

$$Q_\hbar(f_1) \circ Q_\hbar(f_2) = Q_\hbar(f_1 \star_\hbar f_2).$$

In paradigmatic cases, the star product $\star_\hbar$ defines a family of associative algebras $(A_\hbar, \star_\hbar)$, depending on the parameter $\hbar$, and admitting embeddings $C^\infty(M) \hookrightarrow A_\hbar$.

Thinking of $\hbar$ as a deformation parameter, one can interpret $(A_\hbar, \star_\hbar)$ as a (non-commutative) deformation of the usual (commutative) algebra of functions

A. Cabrera
Departamento de Matemática, Instituto de Matemática, Universidade Federal do Rio de Janeiro, Rio de Janeiro, Brazil
e-mail: alejandro@matematica.ufrj.br

R. L. Fernandes (✉)
Department of Mathematics, University of Illinois at Urbana-Champaign, Urbana, IL, USA
e-mail: ruiloja@illinois.edu

M. Abreu et al. (eds.), *Brazil-Portugal Mathematics*, Coimbra Mathematical Texts 4,
https://doi.org/10.1007/978-3-031-98455-6_3

$(C^\infty(M), \cdot)$. At this point one may forget about the original quantization functor $Q_\hbar$ and look for any such deformation with the property that

$$\{f, g\} = \lim_{\hbar \to 0} \frac{1}{\hbar} (f \star_\hbar g - g \star_\hbar f) .$$

This equation is known as the *classical correspondence principle* and one says that $\star_\hbar$ is a deformation of the usual product in the direction of the Poisson bracket $\{\cdot, \cdot\}$.

One way to make this heuristic discussion precise is to consider **formal star products** deforming $\{\cdot, \cdot\}$ which are given by formal power series

$$f_1 \star_\hbar f_2 := \sum_{k=0}^{\infty} B_k(f_1, f_2)\hbar^k ,$$

and satisfy a set of axioms (see Sect. 3). The formality conjecture stated and proved by Kontsevich [20] implies that every Poisson bracket admits a formal star product and that they can be classified.

While formal deformation quantization, after Kontsevich's work, is a remarkably successful theory, it only partially captures the original features that were intended to be made mathematically precise in the transition from classical to quantum mechanics. For example, Kontsevich noted in [20], "It is not clear whether 'deformation quantization' is natural for quantum mechanics (...) A topological open string theory seems to be more relevant." Additionally, Gukov and Witten in [16] emphasize that "(formal) deformation quantization is not quantization." Consequently, there is a significant interest in non-formal deformation quantization, of which much less is known.

A particular case of such non-formal quantizations uses C^*-algebras and is called *strict deformation quantizations* (see, e.g., [25] for a precise formulation). An analogue of Kontsevich's result for strict deformation quantization is not known and Rieffel in [26] poses the following questions.

Fundamental Problem
When does a Poisson manifold admit a strict deformation quantization? In particular, what cohomological obstructions are there to having a strict deformation quantization?

There are several variations on the notion of non-formal deformation quantization—see, e.g., [27] for a recent survey. In recent unpublished work we propose a new approach to non-formal star products $\star_\hbar$. We consider star products given by the sort of *semi-classical Fourier integral operators (SCFIO)* studied in semi-classical analysis—see, e.g., [15, 21, 33]. Two reasons for this approach are:

(i) the need to consider products $\star_\hbar$ defined on $\hbar$-dependent families of smooth functions that may become singular as $\hbar \to 0$, such as highly oscillatory functions $e^{\frac{i}{\hbar}\langle \cdot, \xi_0 \rangle}$ and
(ii) several successful quantization schemes, such as the Weyl quantization of cotangent bundles [15] or Rieffel's quantization of linear Poisson brackets [25], can be expressed in terms of semi-classical Fourier integral operators.

The use of semi-classical analysis then suggests that the axioms of a star product should only hold microlocally. These ideas have many precedents (see [7, 19, 30] and references therein). Also, oscillatory integrals have been considered in the study of strict deformation quantization (see, e.g., [2, 22] and the survey [27]) but, as far as the authors know, ours is the first systematic treatment which incorporates all the semi-classical and microlocal features into play and establishes a detailed relation to the Lie theory of the underlying Poisson manifold.

In this note we sketch our work in this direction, including a provisional definition of non-formal deformation quantization. We describe some conjectural results, attempting to extend results valid in the formal setting to the non-formal setting. We will explain why we believe in the following:

Motto
If a Poisson manifold $(M, \{\cdot, \cdot\})$ admits an *associative non-formal quantization*, then it must be integrable by a symplectic groupoid.

This note is organized as follows. In Sect. 2, we recall some basic facts about Poisson manifolds and symplectic groupoids. In Sect. 3, we review formal deformation quantization. In Sect. 4, we give a tentative definition of non-formal deformation quantization, using SCFIOs, we state some conjectural results and we explain what led us to our motto above.

2 Poisson Geometry

The geometry underlying many fundamental classical physical systems is *Poisson geometry*, whose objects of study are Poisson manifolds, i.e., manifolds equipped with a Poisson bracket. On a Poisson manifold, a choice of a function—the so-called hamiltonian—determines the Hamiltonian dynamics and mechanical observables are also given by functions. Therefore, the ultimate goal is to quantize a Poisson manifold together with all functions of interest. In this section we provide a brief sketch of Poisson geometry and we refer to the monograph [13] for a in-depth introduction to the subject.

2.1 Poisson Manifolds

A **Poisson bracket** on a manifold M is a Lie bracket on the vector space of smooth functions

$$\{\,,\,\} : C^\infty(M) \times C^\infty(M) \to C^\infty(M),$$

which satisfies the Leibniz type identity

$$\{f, f_1 f_2\} = \{f, f_1\} f_2 + f_1 \{f, f_2\}.$$

The pair $(M, \{\cdot,\cdot\})$ is called a **Poisson manifold**.

On a Poisson manifold $(M, \{\cdot,\cdot\})$, to each function $h \in C^\infty(M)$ is assigned a vector field X_h on M by setting

$$X_h(f) := \{h, f\}.$$

One calls h a **hamiltonian function** and X_h the associated **hamiltonian vector field**.

In the language of classical mechanics, the Poisson manifold $(M, \{\cdot,\cdot\})$ is identified with the *phase space* (or "state space") and the equations for the trajectories of the hamiltonian vector field X_h are called the *Hamilton's equations of the motion*. In local coordinates $(x^1, \ldots, x^m)$, the Poisson bracket can be written has

$$\{f_1, f_2\}(x) = \sum_{i,j=1}^{n} \pi^{ij}(x) \frac{\partial f_1}{\partial x^i} \frac{\partial f_2}{\partial x^j}, \tag{1}$$

and a hamiltonian vector field X_h has the expression

$$X_h = \sum_{i,j=1}^{n} \pi^{ij} \frac{\partial h}{\partial x^i} \frac{\partial}{\partial x^j}.$$

Hamilton's equations are then

$$\dot{x}^k = \{h, x^k\} = \sum_{i=1}^{n} \pi^{ik} \frac{\partial h}{\partial x^i}, \quad (k = 1, \ldots, m). \tag{2}$$

Let us illustrate this with several examples.

2.2 Symplectic Poisson Brackets

The motion of a particle $q(t) \in \mathbb{R}^d$ with mass m subject to a conservative force with potential $V : \mathbb{R}^d \to \mathbb{R}$ is described by Newton's equations

$$m\ddot{q}^k(t) = -\frac{\partial V}{\partial q^k}, \quad (k = 1, \dots, d),$$

which can be rewritten as the first order system

$$\begin{cases} \dot{q}^k = \frac{p_k}{m_k} \\ \dot{p}_k = -\frac{\partial V}{\partial q^k} \end{cases}, \quad (k = 1, \dots, d). \tag{3}$$

To interpret it in the language of Poisson geometry, we consider $M = \mathbb{R}^{2d}$ with coordinates (q^k, p_k) and a Poisson bracket defined by

$$\{f_1, f_2\} := \sum_{k=1}^{d} \left(\frac{\partial f_1}{\partial p_k} \frac{\partial f_2}{\partial q^k} - \frac{\partial f_1}{\partial q^k} \frac{\partial f_2}{\partial p_k} \right). \tag{4}$$

Given any function $h : \mathbb{R}^{2n} \to \mathbb{R}$, the corresponding Hamilton's equations (2) become

$$\begin{cases} \dot{q}^k = \frac{\partial h}{\partial p_k} \\ \dot{p}_k = -\frac{\partial h}{\partial q_k} \end{cases}, \quad (k = 1, \dots, d).$$

If we let

$$h = \sum_{k=1}^{d} \frac{p_k^2}{2m_k} + V(q),$$

Hamilton's equations reduce to Newton's equations in the form (3).

The Poisson bracket in this example belongs to an important class of Poisson brackets, called *symplectic*. In general, a symplectic form on a manifold M of even dimension $n = 2d$ is a 2-form $\omega \in \Omega^2(M)$ satisfying

$$\mathrm{d}\omega = 0, \qquad \underbrace{\omega \wedge \cdots \wedge \omega}_{d\text{-times}} \neq 0$$

Given a symplectic form, each choice of a function $f \in C^\infty(M)$ determines a vector field X_f by

$$i_{X_f}\omega = \mathrm{d}f.$$

It is easy to check that the expression

$$\{f, g\}_\omega := \omega(X_f, X_g)$$

defines a Poisson bracket on M. Any Poisson bracket arising in this way is called a symplectic Poisson bracket.

The previous example is symplectic because it arises from the symplectic form on $\mathbb{R}^{2d}$ given by

$$\omega = \sum_{k=1}^{d} \mathrm{d}q^k \wedge \mathrm{d}p_k.$$

Symplectic Poisson brackets can be characterized as follows

Lemma 1 *A Poisson bracket* $\{\cdot,\cdot\}$ *on* M *is symplectic if for any choice of local coordinates* $(x^1,\ldots,x^m)$ *the matrix* $\pi^{ij} = \{x^i, x^j\}$ *is invertible.*

2.3 Linear Poisson Brackets

The motion of a rigid body around its center of gravity with moments of inertia I_1, I_2 and I_3 is described by Euler's equations

$$\begin{cases} \dot{x}^1 = \frac{I_2 - I_3}{I_2 I_3} x^2 x^3 \\ \dot{x}^2 = \frac{I_3 - I_1}{I_3 I_1} x^3 x^1, \\ \dot{x}^3 = \frac{I_1 - I_2}{I_1 I_2} x^1 x^2. \end{cases} \tag{5}$$

One can interpret it in the language of Poisson geometry by introducing a Poisson bracket on $\mathbb{R}^3$ defined by

$$\{f, g\}(x) := (\nabla f(x) \times \nabla g(x)) \cdot x.$$

and considering the hamiltonian function

$$h = \sum_{i=1}^{3} \frac{(x^i)^2}{2I_i}.$$

Then Euler's equations (5) can be written in Hamiltonian form

$$\dot{x}^k = \{h, x^k\}, \quad (k = 1, 2, 3).$$

In general, given a Lie algebra $\mathfrak{g}$ the dual vector space $M = \mathfrak{g}^*$ has a Poisson bracket defined by

$$\{f, g\}(\xi) = \langle [\mathrm{d}_\xi f, \mathrm{d}_\xi g]_{\mathfrak{g}}, \xi \rangle.$$

One recovers the Poisson bracket in the previous example by letting $\mathfrak{g} = \mathfrak{so}(3,\mathbb{R})$, so that

$$M = \mathfrak{so}(3,\mathbb{R})^* \simeq \mathbb{R}^3.$$

These type of Poisson brackets can be characterized as follows.

Lemma 2 *A Poisson bracket* $\{\cdot,\cdot\}$ *on a vector space* V *is of the form* $\mathfrak{g}^*$ *if and only for any pair of linear functions* f_1 *and* f_2*, their Poisson bracket* $\{f_1, f_2\}$ *is a linear function. Equivalently, if in linear coordinates* $(x^1, \ldots, x^n)$ *the bracket is linear:*

$$\pi^{ij}(x) = \{x^i, x^j\}(x) = \sum_k c_k^{ij} x^k.$$

Notice that for a linear Poisson bracket one has $\pi^{ij}(0) = 0$, so it is never symplectic.

2.4 *Quadratic Poisson Structures*

The time evolution of m species in a closed ecological system can often be described by the Lotka-Volterra equations

$$\dot{x}^i = \varepsilon^i x^i + \sum_{j=1}^m a_{ij} x^i x^j, \quad (i = 1, \ldots, m). \tag{6}$$

for some matrix (a_{ij}) and vector ε^i. One calls the system *conservative* if (a_{ij}) is skew-symmetric and there exists a solution $q = (q_1, \ldots, q_n)$ of the linear system

$$\sum_{j=1}^n a_{ji} q^j = \varepsilon^i.$$

A conservative Lotka-Volterra system can be described as a hamiltonian system as follows. On $\mathbb{R}^m$ consider the Poisson bracket

$$\{f_1, f_2\}(x) := \sum_{i<j}^m a_{ij} x^i x^j \tfrac{\partial f_1}{\partial x^i} \tfrac{\partial f_2}{\partial x^j}.$$

Then the hamiltonian function

$$h = \sum_{i=1}^m \left(q^i \log x^i - x^i\right),$$

defines a hamiltonian vector field and the corresponding Hamilton's equations become the Lotka-Volterra equations (6).

The Poisson bracket in this example is *quadratic*. In general, a Poisson bracket on a vector space V is called homogeneous of degree d if for any linear coordinate system $(x^1, \ldots, x^n)$ the functions $\pi^{ij}(x) = \{x^i, x^j\}(x)$ are homogeneous of degree d.

For example, any quadratic Poisson bracket $\{\cdot,\cdot\}_{\mathbb{R}^{n+1}}$ on $\mathbb{R}^{n+1}$ determines a unique Poisson bracket $\{\cdot,\cdot\}_{\mathbb{P}^n}$ on projective space $\mathbb{P}^n$ for which the projection $\pi:\mathbb{R}^{n+1}\setminus\{0\}\to\mathbb{P}^n$ is a **Poisson map**, i.e., satisfies

$$\{f_1,f_2\}_{\mathbb{P}^n}\circ\pi=\{f_1\circ\pi,f_2\circ\pi\}_{\mathbb{R}^{n+1}},$$

for any pair of functions f_1, f_2 on $\mathbb{P}^n$.

2.5 *The Symplectic Foliation*

The geometry underlying a Poisson bracket on a manifold M has the following three distinct, but intertwined, aspects:

- M inherits an underlying (possibly singular) *foliation* $\mathcal{F}$, i.e., a partition into submanifolds. These submanifolds, called leaves, are the equivalence classes for the equivalence relation on M obtained by declaring two points equivalent if there exists some hamiltonian function h for which the vector field X_h has a trajectory passing through those two points.
- Each leaf L has a unique Poisson structure $\{\cdot,\cdot\}_L$ for which the inclusion $i: L\to M$ is a Poisson map. Moreover, $\{\cdot,\cdot\}_L$ is symplectic, so $\mathcal{F}$ is a foliation by *symplectic leaves*.
- For each $x_0\in M$ there are coordinates such that $X_{x^i}(x_0)=0$ and the leaf through x_0 is given by $\{x^1=\cdots=x^d=0\}$. Then one has the *isotropy Lie algebra* at x_0 with basis $\{e_1,\ldots,e_d\}$, and bracket

$$[e_i,e_j]:=\sum_{k=1}^{d}\frac{\partial\pi^{ij}}{\partial x^k}e_k,$$

  where the π^{ij} are the coefficients of the Poisson bracket, as in (1).

Let us look back at the examples we saw before.

Example 3 The (connected) symplectic manifolds are precisely the Poisson manifolds whose symplectic foliation consists of a single leaf. For this reason, the isotropy Lie group/algebra at any point is trivial.

Example 4 A linear Poisson structure on the dual of a Lie algebra $\mathfrak{g}^*$ has symplectic foliation consisting of the coadjoint orbits. These are obtained as the orbits of a canonical (linear) action on $\mathfrak{g}^*$ of the 1-connected Lie group G whose Lie algebra is $\mathfrak{g}$ (the coadjoint action/representation). The isotropy/Lie algebras of $\mathfrak{g}^*$ are the ones obtained from this action.

Example 5 The symplectic foliation of a quadratic Poisson structure on a vector space V depends heavily on the quadratic homogeneous functions π^{ij}. Consider for example the quadratic Poisson structure on $\mathbb{R}^3$ with Poisson bracket defined by

$$\{x, y\} = xy, \quad \{x, z\} = 0, \quad \{y, z\} = yz,$$

has symplectic leaves the connected components of the surfaces $yz = c$, for $c \neq \{0\}$ and the points (zero dimensional leaves) belonging to the union of the planes $y = 0$ and $z = 0$. On the other hand, the quadratic Poisson bracket defined by

$$\{x, y\} = x^2 + y^2, \quad \{x, z\} = \{y, z\} = 0,$$

has symplectic leaves the cylinders $x^2 + y^2 = c$, with $c > 0$, and the points in the z-axis. For a quadratic Poisson bracket the isotropy Lie algebras are all abelian.

2.6 Symplectic Groupoids

The interaction between Poisson geometry, foliation theory and Lie group theory actually goes much deeper, once Lie algebroids/groupoids are brought into the picture.

Recall that groups typically arise as the symmetries of some given object. The concept of a groupoid allows for more general symmetries, acting on a collection of objects M rather than just a single one. An element in a groupoid $g \in \mathcal{G}$ may be pictured as an arrow from a source object $s(g) \in M$ to a target object $t(g) \in M$

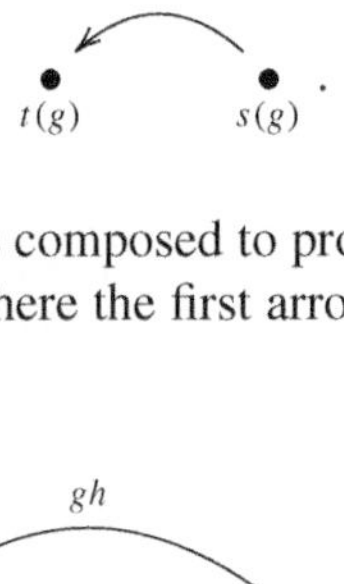

Two such arrows $g, h \in \mathcal{G}$ can be composed to produce a new arrow $gh \in \mathcal{G}$ if and only if the second arrow starts where the first arrow ends, and their product source $s(h)$ and target $t(h)$

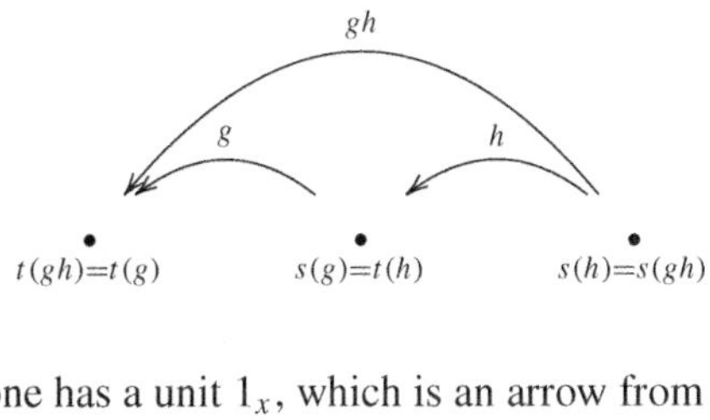

For each object one has a unit 1_x, which is an arrow from x to itself

and for every arrow there is an inverse arrow

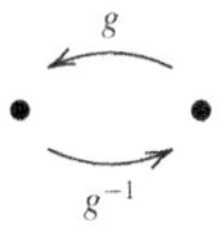

These must satisfy the usual axioms

(i) Associativity: $(gh)k = g(hk)$;
(ii) Identity: $1_{t(g)}g = g1_{s(g)} = g$;
(iii) Inverse: $g^{-1}g = 1_{s(g)}, gg^{-1} = 1_{t(g)}$.

Using categorical language, one can describe a groupoid simply as a category where every arrow has an inverse.

Just as Lie groups—introduced by Lie in the late XIX century—describe smooth symmetries of an object, Lie groupoids—introduced by Ehresmann in the late 1950's—describe smooth symmetries of a smooth family of objects. That is, the collection of arrows is a manifold $\mathcal{G}$, the set of objects is a manifold M, and all the structure maps of the groupoid, namely source, target, multiplication, identity and inverse, are smooth. We denote a groupoid by $\mathcal{G} \rightrightarrows M$.

Given a Lie groupoid $\mathcal{G} \rightrightarrows M$, one has that

- M is partitioned by *orbits* of $\mathcal{G}$, where two objects in M belong to the same orbit if there is an arrow connecting them. The path-connected components of the orbits form a foliation of M;
- the arrows with source and target the same object $x \in M$ form a Lie group $\mathcal{G}_x$, called the *isotropy group* at x.

The groupoids that arise in Poisson geometry are *symplectic groupoids*, i.e., they possess a symplectic form Ω on the space of arrows which is compatible with the groupoid multiplication. One says that a symplectic groupoid $(\mathcal{G}, \Omega) \rightrightarrows M$ integrates a Poisson manifold $(M, \{\cdot, \cdot\})$ it the target map $t : (\mathcal{G}, \{\cdot, \cdot\}_\Omega) \to (M, \{\cdot, \cdot\})$ is a Poisson map. When this happens, the symplectic leaves and the isotropy Lie algebras of $(M, \{\cdot, \cdot\})$ coincide, respectively, with the connected components of the orbits and the Lie algebras of the isotropy groups of $\mathcal{G}$. We refer the reader to [13] for details, and look at the examples above.

Example 6 Given a manifold M the pair groupoid $M \times M \rightrightarrows M$ has arrows the pairs (x, y). Such a pair has source x and target x and the multiplication of two arrows (x, y) and (z, w) can be performed when $y = z$, the resulting arrow being $(x, z) \cdot (z, w) = (z, w)$. When (M, ω) is a symplectic manifold, the pair groupoid inherits a symplectic form, namely $\Omega := \mathrm{pr}_1^* \omega - \mathrm{pr}_2^* \omega$. This form is compatible with multiplication and the resulting symplectic groupoid integrates the Poisson manifold $(M, \{\cdot, \cdot\}_\omega)$.

Example 7 Given an action of a Lie group G on a manifold M one can form the action groupoid $G \times M \rightrightarrows M$, where a pair (g, m) is viewed as an arrow from m to gm. The multiplication of arrows is given by

$$(h, m') \cdot (g, m) = (hg, m), \text{ provided } m' = gm.$$

The linear Poisson structure on $\mathfrak{g}^*$ can be integrated by the action groupoid $G \times \mathfrak{g}^* \rightrightarrows \mathfrak{g}^*$ associated with the coadjoint action of G on $\mathfrak{g}^*$. The symplectic form is obtained from the canonical identification $T^*G \simeq G \times \mathfrak{g}^*$, since any cotangent bundle has a natural symplectic form.

Example 8 Quadratic Poisson brackets also admit integrating symplectic groupoids. The formulas are more complicated, so we consider only the simplest example of the quadratic Poisson bracket on $\mathbb{R}^2$ given by

$$\{x, y\} = xy.$$

Then one finds a symplectic groupoid $(\mathbb{R}^4, \Omega) \rightrightarrows \mathbb{R}^2$, where the source and target maps are

$$s(x, y, u, v) = (x, y), \quad t(x, y, u, v) = (xe^{yv}, ye^{-xu}),$$

and the multiplication is given by the formula

$$(x', y', u', v') \cdot (x, y, u, v) = (x, y, u + e^{yv}u', v + e^{-xu}v').$$

The relevant symplectic form is:

$$\Omega = -\mathrm{d}(xu) \wedge \mathrm{d}(yv) + \mathrm{d}x \wedge \mathrm{d}u + \mathrm{d}y \wedge \mathrm{d}v.$$

Given a Poisson manifold $(M, \{\cdot, \cdot\})$ one may ask if there is some symplectic groupoid integrating $(M, \{\cdot, \cdot\})$. In the three examples above the Poisson manifolds are all integrable, but this does not need to be the case. The obstructions to integrability are well understood—see [10–12].

In order to study global properties of Poisson manifolds and their symplectic foliations, one must pay attention to their integrating symplectic groupoids. We will see that symplectic groupoids are also central ingredients in quantization schemes for Poisson manifolds, both at formal and non-formal levels [3, 17, 18] and naturally arise in connection with topological field theories (the Poisson sigma model) [8].

3 Formal Deformation Quantization

From now it is convenient to work with $\mathbb{C}$-valued functions, so $C^\infty(M)$ will denote the space of $\mathbb{C}$-valued smooth functions and we consider Poisson brackets on such functions.

3.1 Formal Star Products

A **formal star product** for a general Poisson manifold $(M, \{\cdot, \cdot\})$ is a binary operation on the space of formal power series $C^\infty(M)[[\hbar]]$, defined for a pair of functions $f_1, f_2 \in C^\infty(M)$ by an expression of the form

$$f_1 \star_\hbar f_2 := \sum_{k=0}^{\infty} B_k(f_1, f_2)\hbar^k, \tag{7}$$

and extended to the space of formal power series by requiring $\mathbb{C}[[\hbar]]$-linearity. On also imposes the following axioms:

(A1) *Deformation of usual product.* For every $f_1, f_2 \in C^\infty(M)$,

$$f_1 \star_\hbar f_2 = f_1 f_2 + O(\hbar);$$

(A2) *Correspondence principle.* For every $f_1, f_2 \in C^\infty(M)$,

$$\{f_1, f_2\} = \frac{1}{i\hbar}(f_1 \star_\hbar f_2 - f_2 \star_\hbar f_1) + O(h);$$

(A3) *Associativity.* For every $f_1, f_2, f_3 \in C^\infty(M)[[\hbar]]$,

$$(f_1 \star_\hbar f_2) \star_\hbar f_3 = f_1 \star_\hbar (f_2 \star_\hbar f_3);$$

(A4) *Identity.* For every $f \in C^\infty(M)[[\hbar]]$,

$$1 \star_\hbar f = f = f \star_\hbar 1.$$

One calls $\star_\hbar$ *natural* if the B_k are bidifferential operators of order $\leq k$ in each entry.

The formality conjecture stated and proved by Kontsevich [20] yields the following fundamental result.

Theorem 9 (Kontsevich) *Every Poisson manifold admits a natural formal star product.*

Kontsevich actually gives a classification of star products for a given Poisson manifold and a beautiful explicit formula to construct a star product for any Poisson structure in $\mathbb{R}^n$.

3.2 Examples of Formal Star Products

In general, in spite of Kontsevich formula, it is hard to give explicit examples of star products deforming a given Poisson bracket. Still, in some cases, explicit formulas can be given and we list here a few examples.

Example 10 On $M = \mathbb{R}^m$ any constant skew-symmetric matrix (π^{ij}) defines a constant Poisson structure

$$\{x^i, x^j\} := \pi^{ij}.$$

A formal star product deforming this Poisson bracket is given by the Moyal product

$$f_1 \star_\hbar f_2 := \exp\left(\frac{i\hbar}{2}\pi^{ij}\frac{\partial}{\partial x^i}\frac{\partial}{\partial y^j}\right) f_1(x) f_2(y)\Big|_{y=x} \tag{8}$$

$$= \sum_{k=0}^{\infty} \frac{1}{k!}\left(\frac{i\hbar}{2}\right)^k \pi^{i_1 j_1} \cdots \pi^{i_k j_k} \frac{\partial^k f_1}{\partial x^{i_1} \cdots \partial x^{i_k}} \frac{\partial^k f_2}{\partial x^{j_1} \cdots \partial x^{j_k}},$$

For example, this formula gives a formal star product deforming the symplectic Poisson bracket on $\mathbb{R}^{2d}$ defined by (4).

Example 11 Let $M = \mathfrak{g}^*$ with its linear Poisson structure. To obtain a star product deforming it one can use the Poincaré-Birkoff-Witt isomorphism between the symmetric and the universal enveloping algebras of $\mathfrak{g}$

$$q : S(\mathfrak{g}) \to U(\mathfrak{g}).$$

The symmetric algebra $S(\mathfrak{g})$ is naturally isomorphic to the algebra of polynomial functions on $\mathfrak{g}^*$. Then one defines a $\star_\hbar$-product on polynomials by first defining it on homogeneous polynomials $f_1 \in S^k(\mathfrak{g})$ and $f_2 \in S^l(\mathfrak{g})$ by the formula

$$f_1 \star_\hbar f_2 := \sum_{n=0}^{k+l} (i\hbar)^n \, \mathrm{pr}_{k+l-n}(q^{-1}(q(f_1) \cdot q(f_2))),$$

where $\mathrm{pr}_k : S(\mathfrak{g}) \to S^k(\mathfrak{g})$ denotes the projection onto the homogenous degree k polynomials and $\cdot$ denotes the product in $U(\mathfrak{g})$. The $\mathbb{C}[[\hbar]]$-extension of this map defines a star product on $S(\mathfrak{g})[[\hbar]]$.

This star product is given by a series of the form (7) where the B_k are bidifferential operators, so it also defines a star product on $C^\infty(\mathfrak{g}^*)[[\hbar]]$. Note that if $\{x_1, \ldots, x_m\} \subset \mathfrak{g} = S^1(\mathfrak{g})$ is a basis then the formula above yields

$$x_i \star_\hbar x_j = x_i x_j + \frac{i\hbar}{2}[x_i, x_j].$$

so this star product deforms the linear Poisson structure on $\mathfrak{g}^*$.

Example 12 On $M = \mathbb{R}^m$ any skew-symmetric matrix (a^{ij}) defines a "diagonal" quadratic Poisson bracket by

$$\{x^i, x^j\} := a^{ij} x^i x^j,$$

A star product deforming this Poisson bracket is also given by a Moyal type formula (8) where now $\pi^{ij} = a^{ij} x^i x^j$ is not constant anymore. One explanation for this coincidence is that the change of coordinates $u^i = \log x^i$ transforms this Poisson structure into a constant Poisson structure (but note this is only valid in the domain $x^i > 0$). If one considers a non-diagonal Poisson structure, such as the second Poisson bracket in Example 5, their star products take a more complicated form.

3.3 *Formal Symplectic Groupoid*

Given a Poisson manifold, a symplectic groupoid integrating it can be thought of as a first step towards the quantization of the Poisson bracket. In fact, symplectic groupoids were introduced by Karasev [19], Weinstein [28] and Zakrzewski [31, 32] precisely with the aim of quantizing Poisson manifolds. This quantization program was never completed. However, in the formal setting Cattaneo-Dherin-Felder [3] and Karabegov [18] found the following precise connection between groupoids and quantization.

Theorem 13 *A natural formal $\star_\hbar$-product deforming a Poisson manifold $(M, \{\cdot, \cdot\})$ determines a formal symplectic groupoid structure on a formal neighborhood of the zero section of T^*M which integrates the Poisson bracket.*

Let us briefly explained the concepts from formal geometry used in the statement of this result. Given a closed submanifold $M \subset N$, denote by $I(M) \subset C^\infty(N)$ the ideal formed by the functions that vanish on M. Letting $I^\infty(M) = \bigcap_{k=1}^\infty I(M)^k$, the quotient algebra

$$C^\infty_M(N) := C^\infty(N)/I^\infty(M)$$

is viewed as the algebra of functions on a formal neighborhood of M in N.

In order to define a groupoid structure on a formal neighborhood of the zero section of T^*M, one reinterprets the structure maps in a groupoid $\mathcal{G} \rightrightarrows M$ as pullback maps. For example, pullback by source and target are algebra maps $s^*, t^* : C^\infty(M) \to C^\infty(\mathcal{G})$, and so in the formal setting they are replaced by algebras maps

$$s^*, t^* : C^\infty(M) \to C^\infty_M(T^*M).$$

Similarly, in the formal setting, the inverse map is replaced by the algebra map

$$i^* : C^\infty_M(T^*M) \to C^\infty_M(T^*M),$$

while the unit map $u : M \to \mathcal{G}$ becomes the evaluation map

$$u^* : C^\infty_M(T^*M) \to C^\infty(M).$$

In order to define the multiplication observe that for a groupoid this is a map $m : \mathcal{G}^{(2)} \to \mathcal{G}$ where the domain is the space of composable arrows

$$\mathcal{G}^{(2)} := \{(g, h) \in \mathcal{G} \times \mathcal{G} : s(g) = t(h)\}.$$

One can identify the formal neighborhood of the diagonal $\Delta_M \subset \mathcal{G}^{(2)}$ using the short exact sequence of algebras

$$0 \longrightarrow I_2 \longrightarrow C^\infty_{M\times M}(\mathcal{G} \times \mathcal{G}) \longrightarrow C^\infty_{\Delta_M}(\mathcal{G}^{(2)}) \longrightarrow 0$$

where I_2 is the ideal generated by the image of the map

$$(\mathrm{id} \otimes s^* - t^* \otimes \mathrm{id}) : C^\infty(M) \to C^\infty_{M\times M}(\mathcal{G} \times \mathcal{G}).$$

Hence, in the formal setting, multiplication is defined as an algebra map

$$m^* : C^\infty_M(T^*M)^{(2)} \to C^\infty_M(T^*M),$$

where by definition

$$C^\infty_M(T^*M)^{(2)} := C^\infty_{M\times M}(T^*M \times T^*M)/I_2.$$

Similarly, one translates the axioms that the structure maps must satisfy in terms of pullback diagrams—see [18].

Finally, the ideal $I^\infty(M)$ is also a Poisson ideal for the Poisson bracket $\{\cdot, \cdot\}_{\mathrm{can}}$ associated with the canonical symplectic form ω_{can} on T^*M. Hence, one obtains a Poisson bracket on the algebra of formal functions $C^\infty_M(T^*M)$. The algebra $C^\infty_M(T^*M)^{(2)}$ also inherits a Poisson bracket, and so one defines a **formal symplectic groupoid** integrating a Poisson manifold $(M, \{\cdot, \cdot\})$, to consist of:

- a formal groupoid structure in the formal neighborhood of the zero section $0 \subset T^*M$, such that
- multiplication $m^* : (C^\infty_M(T^*M)^{(2)}, \{\cdot, \cdot\}_{\mathrm{can}}) \to (C^\infty(T^*M, M, \{\cdot, \cdot\}_{\mathrm{can}})$ and target $t^* : (C^\infty(M), \{\cdot, \cdot\}) \to (C^\infty_M(T^*M), \{\cdot, \cdot\}_{\mathrm{can}})$ are Poisson maps.

Given a Poisson manifold there always exist a *local* symplectic groupoid integrating it. This is a classical result due to Coste, Dazord and Weinstein [9]. From this it follows also that there exists always a formal symplectic groupoid integrating the Poisson manifold. This can also be seen from quantization via Theorem 13 since, by Kontsevich, any Poisson manifold admits a formal deformation quantization.

4 Non-formal Deformation Quantization

4.1 Definition and First Examples

We propose to study non-formal star products $\star_\hbar$ given by the sort of semi-classical Fourier integral operators (SCFIO) studied in semi-classical analysis. In this approach a tentative definition go as follows, where we assume for simplicity that M is compact.

A **non-formal deformation quantization** of a Poisson manifold $(M, \{\cdot, \cdot\})$ consists of a bilinear operation $\star_\hbar$ of the form

$$f_1 \star_\hbar f_2 := F_\hbar(f_1 \otimes f_2),$$

where $F_\hbar$ is a SCFIO of order $k = \frac{\dim M}{4}$, satisfying the following axioms:

(A1) *Deformation of usual product:* For any $f_1, f_2 \in C^\infty(M)$:

$$f_1 \star_\hbar f_2 - f_1 f_2 = O(\hbar).$$

(A2) *Correspondence principle:* For any $f_1, f_2 \in C^\infty(M)$:

$$\{f_1, f_2\} - \frac{1}{i\hbar}(f_1 \star_\hbar f_2 - f_2 \star_\hbar f_1) = O(\hbar).$$

(A3) *Associativity:* For any h-dependent families f_1, f_2 and f_3, one has

$$(f_1 \star_\hbar f_2) \star_\hbar f_3 - f_1 \star_\hbar (f_2 \star_\hbar f_3) = O(\hbar^\infty).$$

(A4) *Identity:* For any h-dependent family f:

$$(1 \star_\hbar f - f) = O(\hbar), \quad (f \star_\hbar 1 - f)|_U = O(\hbar).$$

Note that the notation $O(\hbar)$ in the previous axioms has a different meaning from the same notation used in the formal case, in Sect. 3.1. Here $O(\hbar)$ and $O(\hbar^\infty)$ should be taken for norms in appropriate function spaces. Also, in the last two axioms the equality holds microlocally in a neighborhood of the zero section.

We will say that such non-formal deformation quantization is *natural* when the left and right translation operators, $L_f(g) := f \star_\hbar g$ and $R_f(g) := g \star_\hbar f$, are (semi-classical) pseudo-differential operators of order zero on M, microlocally near the zero section. Our assumption about the order of the SCFIO $F_\hbar$ in the definition above is necessary for L_f and R_f to have order 0. We shall only consider non-formal deformation quantizations which are natural.

Underlying any SCFIO $F_\hbar$, and hence also a non-formal deformation quantization, one has canonical relation $\Lambda : M \times M \dashrightarrow M$, i.e., a Lagrangian submanifold

$$\Lambda \subset \overline{T^*M \times T^*M} \times T^*M.$$

Let us give two paradigmatic examples.

Example 14 The very first example arises, of course, from the Weyl quantization of the canonical Poisson bracket on $M = \mathbb{R}^{2n}$. The corresponding star products can be expressed as SCFIO by

$$f_1 \star_\hbar f_2(x, p) := \frac{1}{(\pi\hbar)^{2d}} \int_{\alpha_1, \alpha_2 \in \mathbb{R}^{2d}} f_1(\alpha_1) f_2(\alpha_2) e^{\frac{i}{\hbar} S} \, d\alpha_1 d\alpha_2,$$

where $\alpha = (x, p)$ and

$$S(\alpha, \alpha_1, \alpha_2) := \int_{\Delta(\alpha, \alpha_1, \alpha_2)} 4\omega_{\mathrm{can}}.$$

with $\Delta(\alpha, \alpha_1, \alpha_2) \subset \mathbb{R}^{2d}$ the euclidean triangle with vertices α, α_1 and α_2.

Example 15 A more instructive example, where one already sees the need to work microlocally, is given by the linear Poisson structure on the dual of a Lie algebra $M = \mathfrak{g}^*$. Following an idea of Rieffel [24], one can proceed as follows. Let $0 \in U \subset \mathfrak{g}$ be an open set where the exponential map $\exp : \mathfrak{g} \to G$ restricts to a diffeomorphism onto its image, so U becomes a local Lie group with product denoted $\bullet$. We then choose open balls $D \subsetneq D' \subset U$ centered at 0 such that $D^3 \subset U$, $(D')^2 \subset U$. Also, we choose a Haar density on G and transport it to a density μ on U using the exponential map. After a possible rescaling, one find that:

$$\mu(p) = \det\left(\frac{I - e^{-\mathrm{ad}_p}}{\mathrm{ad}_p}\right) dp,$$

where dp denotes the Lebesgue density on $\mathfrak{g}$. Then we set:

$$f_1 \star_\hbar f_2 := \frac{1}{(2\pi\hbar)^{n/2}} \mathfrak{F}_\hbar^{-1}[(\rho\mathfrak{F}_\hbar(f_1)) \cdot_U (\rho\mathfrak{F}_\hbar(f_2))],$$

where $\mathfrak{F}_\hbar$ denotes the semi-classical Fourier transform, $\rho \in C_c^\infty(U)$ is a cutoff function with $\rho|_D \equiv 1$, and $\cdot_U$ denotes the convolution in the local group U with respect to μ. Explicitly, we find the oscillatory integral expression

$$f_1 \star_\hbar f_2(x) = \frac{1}{(2\pi\hbar)^{n/2}} \int_{x_1, x_2} \int_{p, \tilde{p} \in U} f_1(x_1) f_2(x_2) e^{\frac{i}{\hbar} S} \rho(\tilde{p}) \rho(\tilde{p}^{-1} \bullet p) \mu(\tilde{p}) dx_1 dx_2,$$

where:

$$S(p, x, x_1, x_2) := px - \tilde{p}x_1 - (\tilde{p}^{-1} \bullet p)x_2.$$

If these expressions did not include a cutoff function ρ, we would need the functions f_i to have Fourier transform $\mathfrak{F}_\hbar(f_i)$ with support contained in D so the expressions make sense. This would allow to have the axioms for $\star_\hbar$ be satisfied on the nose, but would greatly limit the domain of $\star_\hbar$. The above choice which includes the cutoff makes the domain bigger while the axioms for $\star_\hbar$ only hold microlocally.

There are other examples of Poisson manifolds that admit these type of star products. In general, the question of which Poisson maniifolds admit such non-formal deformation quantization is widely open.

Our first main (conjectural) result concerns the Lagrangian submanifold $\Lambda \subset T^*(M \times M \times M)$ underlying the SCFIO defining $\star_\hbar$ in the definition above.

Theorem 16 (conjecture) *If $\star_\hbar$ is a non-formal star product deforming a Poisson manifold (M, π) then Λ is the graph of multiplication of a local symplectic groupoid integrating (M, π).*

This result is the non-formal analogue of Theorem 13. The proof should use the calculus of semi-classical wavefront sets as developed in [1] and the symbol calculus for semi-classical pseudo-differential operators (see, e.g., [15]) applied to the left and right translation operators, L_f and R_f.

Our main motivation to propose this semi-classical approach is that it is general enough to properly quantize many Poisson manifolds. Still, we also expect it to be related to formal deformation quantization as follows.

Theorem 17 (conjecture) *Let $\star_\hbar$ is a non-formal star product deforming a Poisson manifold (M, π). For every $x_0 \in M$ and every $f, g \in C_c^\infty(M)$ with small enough support around x_0, $f \star_\hbar g$ extends smoothly to $\hbar = 0$ and its Taylor series at $\hbar = 0$*

$$\mathcal{T}_\epsilon(f \star_\hbar g(x_0)) = \sum_{n \geq 0} \epsilon^n \; B_n(f, g)|_{x_0} \tag{9}$$

defines a natural formal star product $\star_\epsilon$ on $C^\infty(M)[[\epsilon]]$ quantizing (M, π).

The proof of this result should follow by using the calculus of wavefront sets and the law of composition of SCFIOs, to conclude that $f \star_\hbar g$ is of degree 0 and hence extends smoothly to $\hbar = 0$. The axioms (A1)–(A4) in definition will then imply that (9) satisfies the axioms of a formal start product.

The definition of non-formal start product $\star_\hbar$ above also yields a *partial algebra* $(\mathcal{A}_W, \star_\hbar)$, where $W \subset T^*M$ is a neighborhood of 0_M and

$$\mathcal{A}_W := \{f : \mathrm{WF}_\hbar(f) \subset W\}/O(\hbar^\infty).$$

Here $\mathrm{WF}_\hbar(f)$ denotes the semi-classical wave front set of the $\hbar$-dependent family f (see [33]). We abuse notation and we still use the symbol $\star_\hbar$ for multiplication of equivalence classes in $\mathcal{A}_W$. Appealing to a microlocal continuity property, it is a partially defined product with domain:

$$\mathrm{Dom}(\mathcal{A}_W, \star_\hbar) := \{(f, g) : \mathrm{WF}_\hbar(f \star_\hbar g) \subset W\}.$$

We say that $\star_\hbar$ is a *strict non-formal star product* if there exists W for which $\mathcal{A}_W$ embeds in an algebra. Our third main (conjectural) result is the following theorem relating existence of strict deformation quantizations and integrability of a Poisson structure:

Theorem 18 (conjectural) *If (M, π) admits a strict non-formal deformation quantization then it is integrable by a symplectic groupoid.*

In order to explain why this should hold, recall that in a partial algebra the associativity property reads:

$$a_1 \cdot (a_2 \cdot a_3) = (a_1 \cdot a_2) \cdot a_3, \quad \textit{provided both sides are defined.}$$

This *does not* imply that higher associativities hold. For example, consider all possible products of 4 elements:

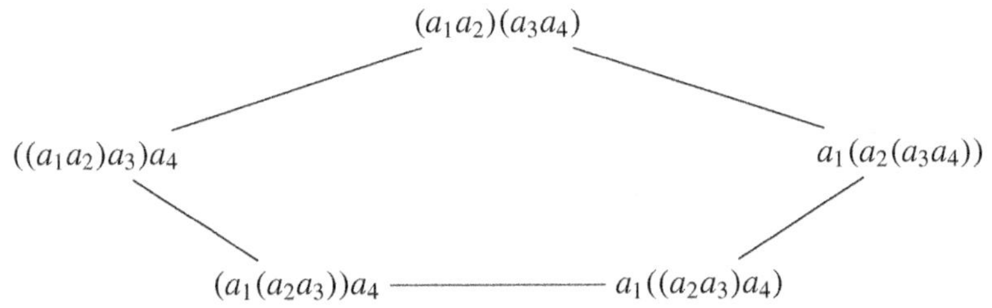

Each edge represents a move that uses only the 3-associativity property. In an algebra all vertices of the pentagon are defined and the 3-associativity implies 4-associativity. In a partial algebra it is possible that, for example, $(a_1a_2)(a_3a_4)$ and $(a_1(a_2a_3))a_4$ are defined but none of the other vertices are defined, and then 3-associativity does not allow one to conclude that 4-associativity holds. For a partial algebra to embed in an algebra n-associativity must hold for all n, in which case one says that the algebra is *globally associative*.

A similar situation happens with local Lie groups and local Lie groupoids. A classical theorem of Mal'cev essentially says that a local Lie group embedds in a global Lie group if and only if the multiplication is globally associative (see [23]). This results was generalized to Lie groupoids in [14] and for the proof of Theorem 18 one needs to show that if $\star_\hbar$ is a strict non-formal star product, then the local symplectic groupoid given by Theorem 16 is globally associative. In [14] the precise relationship between the failure of n-associativity and the obstructions to integrability found in [10, 11] is also established. From it, one should also be able to obtain the following answer to Rieffel's question mentioned in the Introduction:

Conjecture
If (M, π) admits a *strict deformation quantization*, then it must be integrable by a symplectic groupoid. In particular, the obstruction theory of [10, 11] applies.

Since Rieffel's strict deformation quantization is by C^*-algebras, one needs to investigate the relationship with the notion of non-formal deformation quantization above.

Our tentative results also place our theory right at the cross paths of older attempts by Maslov and Karasëv [19], Weinstein [29] and Zakrzewski [31, 32], to construct non-formal deformation quantizations by using symplectic groupoids. In this direction, Hawkins [17] proposed that a Poisson manifold may be quantized by a twisted polarized convolution C^*-algebra of a symplectic groupoid. The most important remaining problem is, as already, mentioned

Open Problem
How can one construct a non-formal deformation quantization of an integrable Poisson manifold by a star product of SCFIO type?

In a series of works, Cattaneo, Dherin and Weinstein [4–7] also use a semi-classical approach, albeit different from ours (e.g., in [7, Sect. 5.7], the domain of $\star_\hbar$ does not include oscillatory functions and the canonical relations underlying SCFIOs must satisfy extra properties from the outset). Their techniques should be very helpful to make progress in this problem.

Acknowledgements AC was partially supported by the grants FAPERJ JCNE E26/203.262/2017 and CNPq 309847/2021-4.

RLF was partially supported by NSF grants DMS-2003223 and DMS-2303586.

References

1. Alexandrova, I.: Semi-classical wavefront set and Fourier integral operators. Canad. J. Math. **60**(2), 241–263 (2008). https://doi.org/10.4153/CJM-2008-011-7
2. Bieliavsky, P.: Strict quantization of solvable symmetric spaces. J. Symplectic Geom. **1**(2), 269–320 (2002). https://doi.org/10.4310/jsg.2001.v1.n2.a4
3. Cattaneo, A.S., Dherin, B., Felder, G.: Formal symplectic groupoid. Comm. Math. Phys. **253**(3), 645–674 (2005). https://doi.org/10.1007/s00220-004-1199-z
4. Cattaneo, A.S., Dherin, B., Weinstein, A.: Symplectic microgeometry I: micromorphisms. J. Symplectic Geom. **8**(2), 205–223 (2010). https://doi.org/10.4310/jsg.2010.v8.n2.a4
5. Cattaneo, A.S., Dherin, B., Weinstein, A.: Symplectic microgeometry II: generating functions. Bull. Braz. Math. Soc. (N.S.) **42**(4), 507–536 (2011). https://doi.org/10.1007/s00574-011-0027-2

6. Cattaneo, A.S., Dherin, B., Weinstein, A.: Symplectic microgeometry III: Monoids. J. Symplectic Geom. **11**(3), 319–341 (2013). https://doi.org/10.4310/jsg.2013.v11.n3.a1
7. Cattaneo, A.S., Dherin, B., Weinstein, A.: Symplectic microgeometry, IV: quantization. Pacific J. Math. **312**(2), 355–399 (2021). https://doi.org/10.2140/pjm.2021.312.355
8. Cattaneo, A.S., Felder, G.: Poisson sigma models and symplectic groupoids. In: Landsman, N.P., Pflaum, M., Schlichenmaier, M. (eds.) Quantization of Singular Symplectic Quotients, Progr. Math., vol. 198, pp. 61–93. Birkhäuser, Basel (2001). https://doi.org/10.1007/978-3-0348-8364-1
9. Coste, A., Dazord, P., Weinstein, A.: Groupoïdes symplectiques. In: Publications du Département de Mathématiques. Nouvelle Série. A, Vol. 2, Publ. Dép. Math. Nouvelle Sér. A, vol. 87, pp. i–ii, 1–62. Univ. Claude-Bernard, Lyon (1987). https://eudml.org/doc/273582
10. Crainic, M., Fernandes, R.L.: Integrability of Lie brackets. Ann. Math. (2) **157**(2), 575–620 (2003). https://doi.org/10.4007/annals.2003.157.575
11. Crainic, M., Fernandes, R.L.: Integrability of Poisson brackets. J. Differ. Geom. **66**(1), 71–137 (2004). http://projecteuclid.org/euclid.jdg/1090415030
12. Crainic, M., Fernandes, R.L.: Lectures on integrability of Lie brackets. In: T. Ratiu, A. Weinstein, N.T. Zung (eds.) Lectures on Poisson Geometry, Geom. Topol. Monogr., vol. 17, pp. 1–107. Geom. Topol. Publ., Coventry (2011). https://msp.org/gtm/2011/17/
13. Crainic, M., Fernandes, R.L., Mărcuţ, I.: Lectures on Poisson Geometry, Graduate Studies in Mathematics, vol. 217. American Mathematical Society, Providence, RI (2021). https://doi.org/10.1090/gsm/217
14. Fernandes, R.L., Michiels, D.: Associativity and integrability. Trans. Amer. Math. Soc. **373**(7), 5057–5110 (2020). https://doi.org/10.1090/tran/8073
15. Guillemin, V., Sternberg, S.: Semi-classical Analysis. International Press, Boston, MA (2013)
16. Gukov, S., Witten, E.: Branes and quantization. Adv. Theor. Math. Phys. **13**(5), 1445–1518 (2009). https://doi.org/10.4310/atmp.2009.v13.n5.a5
17. Hawkins, E.: A groupoid approach to quantization. J. Symplectic Geom. **6**(1), 61–125 (2008). https://doi.org/10.4310/jsg.2008.v6.n1.a4
18. Karabegov, A.V.: Formal symplectic groupoid of a deformation quantization. Comm. Math. Phys. **258**(1), 223–256 (2005). https://doi.org/10.1007/s00220-005-1336-3
19. Karasëv, M.V., Maslov, V.P.: Nonlinear Poisson brackets, Translations of Mathematical Monographs, vol. 119. American Mathematical Society, Providence, RI (1993). https://doi.org/10.1090/mmono/119. Geometry and quantization, Translated from the Russian by A. Sossinsky [A. B. Sosinskiĭ] and M. Shishkova
20. Kontsevich, M.: Deformation quantization of Poisson manifolds. Lett. Math. Phys. **66**(3), 157–216 (2003). https://doi.org/10.1023/B:MATH.0000027508.00421.bf
21. Martinez, A.: An introduction to semiclassical and microlocal analysis. Universitext. Springer, New York (2002). https://doi.org/10.1007/978-1-4757-4495-8
22. Natsume, T., Nest, R., Peter, I.: Strict quantizations of symplectic manifolds. Lett. Math. Phys. **66**(1-2), 73–89 (2003). https://doi.org/10.1023/B:MATH.0000017652.90999.d8
23. Olver, P.J.: Non-associative local Lie groups. J. Lie Theory **6**(1), 23–51 (1996)
24. Rieffel, M.A.: Lie group convolution algebras as deformation quantizations of linear Poisson structures. Amer. J. Math. **112**(4), 657–685 (1990). https://doi.org/10.2307/2374874
25. Rieffel, M.A.: Quantization and C^*-algebras. In: Doran, R.S. (ed.) C^*-algebras: 1943–1993 (San Antonio, TX, 1993), Contemp. Math., vol. 167, pp. 66–97. Amer. Math. Soc., Providence, RI (1994). https://doi.org/10.1090/conm/167/1292010
26. Rieffel, M.A.: Questions on quantization. In: Ge, L., Lin, H., Ruan, Z.J., Zhang, D., Zhang, S. (eds.) Operator Algebras and Operator Theory (Shanghai, 1997), Contemp. Math., vol. 228, pp. 315–326. Amer. Math. Soc., Providence, RI (1998). https://doi.org/10.1090/conm/228/03294
27. Waldmann, S.: Convergence of star products: from examples to a general framework. EMS Surv. Math. Sci. **6**(1-2), 1–31 (2019). https://doi.org/10.4171/emss/31
28. Weinstein, A.: Symplectic groupoids and Poisson manifolds. Bull. Amer. Math. Soc. (N.S.) **16**(1), 101–104 (1987). https://doi.org/10.1090/S0273-0979-1987-15473-5

29. Weinstein, A.: Noncommutative geometry and geometric quantization. In: Donato, P., Duval, C., Elhadad, J., Tuynman, G.M. (eds.) Symplectic Geometry and Mathematical Physics (Aix-en-Provence, 1990), Progr. Math., vol. 99, pp. 446–461. Birkhäuser Boston, Boston, MA (1991)
30. Weinstein, A.: Traces and triangles in symmetric symplectic spaces. In: Maeda, Y., Omori, H., Weinstein, A. (eds.) Symplectic Geometry and Quantization (Sanda and Yokohama, 1993), Contemp. Math., vol. 179, pp. 261–270. Amer. Math. Soc., Providence, RI (1994). https://doi.org/10.1090/conm/179/01945
31. Zakrzewski, S.: Quantum and classical pseudogroups. I. Union pseudogroups and their quantization. Comm. Math. Phys. **134**(2), 347–370 (1990). http://projecteuclid.org/euclid.cmp/1104201734
32. Zakrzewski, S.: Quantum and classical pseudogroups. II. Differential and symplectic pseudogroups. Comm. Math. Phys. **134**(2), 371–395 (1990). http://projecteuclid.org/euclid.cmp/1104201735
33. Zworski, M.: Semiclassical analysis, Graduate Studies in Mathematics, vol. 138. American Mathematical Society, Providence, RI (2012). https://doi.org/10.1090/gsm/138

Linear and Nonlinear Fractional PDEs from Interacting Particle Systems

Pedro Cardoso and Patrícia Gonçalves

Abstract In these notes, we describe the strategy for the derivation of the hydrodynamic limit for a family of long range interacting particle systems of exclusion type with symmetric rates. For $m \in \mathbb{N} := \{1, 2, \ldots\}$ fixed, the hydrodynamic equation is $\partial_t \rho(t, u) = [-(-\Delta)^{\gamma/2}\rho^m](t, u)$. For $m = 1$, this is the fractional equation, which is linear. On the other hand, for $m \geq 2$, this is the fractional porous medium equation (which is nonlinear), obtained by choosing a rate which depends on the number of particles next to the initial and final position of a jump.

1 Introduction

One important problem in statistical mechanics consists in deriving the macroscopic evolution equations of the thermodynamic quantity (ies) of a gas from the interaction of its constituent models. According to Boltzmann, one should first find the equilibrium states of the system and then analyse the evolution out of the equilibrium. If we assume that the molecules perform continuous time random walks with certain constrains, then the evolution equations can be deduced for some dynamics. To answer this question, in the seventies, Spitzer in [16] introduced stochastic interacting particle systems (SIPS) as toy models for a variety of phenomena.

Over the last fours decades, there has been remarkable progress in deriving those macroscopic equations by means of rigorous mathematical results. The way to achieve it is to make a connection through a scaling parameter that links the continuous space where the solutions will be defined and the discrete space where the particles will evolve. In this framework, many partial differential equations (PDEs)

P. Cardoso
Institute for Applied Mathematics, University of Bonn, Bonn, Germany
e-mail: pgondimc@uni-bonn.de

P. Gonçalves (✉)
Center for Mathematical Analysis, Geometry and Dynamical Systems, Instituto Superior Técnico, Universidade de Lisboa, Lisboa, Portugal
e-mail: pgoncalves@tecnico.ulisboa.pt

M. Abreu et al. (eds.), *Brazil-Portugal Mathematics*, Coimbra Mathematical Texts 4,
https://doi.org/10.1007/978-3-031-98455-6_4

have been studied and derived from several underlying SIPS. The list of PDEs is quite vast and the same PDE can be obtained from different macroscopic dynamics. Nevertheless, the nature of these equations highly depends on the chosen microscopic stochastic dynamics: it can be parabolic, hyperbolic, or even of fractional form.

In these notes we focus on the derivation of fractional nonlinear PDEs, though we also discuss the fractional heat equation. More precisely, we are particularly interested in the derivation of the fractional porous medium equation (FPME) given for $\gamma \in (0, 2)$ and $2 \leq m \in \mathbb{N}$ by

$$\begin{cases} \partial_t \rho(t, u) = [-(-\Delta)^{\gamma/2} \rho^m](t, u), \quad u \in \mathbb{R}, \; t \in [0, T], \\ \rho(0, u) = g(u), \quad u \in \mathbb{R}, \end{cases} \tag{1}$$

where $g : \mathbb{R} \to [0, 1]$ is a measurable function. Above, $-(-\Delta)^{\gamma/2}$ denotes the fractional Laplacian operator, which, contrarily to the usual Laplacian operator, is non-local; see (7) for its definition. The diffusive porous medium equation (DPME) is given by$\partial_t \rho(t, u) = \Delta \rho^m(t, u)$ where $m \geq 2$ and models the diffusion of a substance through a porous medium. The quantity $\rho(t, u)$ denotes the density of the substance at time t and position u. Writing the equation in divergence form $\partial_t \rho(t, u) = \nabla(D(\rho)\nabla \rho(t, u))$, the diffusion coefficient is $D(\rho) = m\rho^{m-1}$.

Since the diffusion coefficient $D(\rho)$ vanishes as the density tends to zero, this suggests that at the microscopic level, when the density of particles is low, it becomes more difficult for particles to diffuse. With this scenario in mind, in [11] it was proposed a model of interacting particles that has a hydrodynamic limit given by the DPME, without boundary conditions. The power m derived in that article is an integer number since it is related to the required number of particles close to the sites where an exchange occurs: when m increases, the same occurs with the necessary quantity of particles. The DPME was also obtained in [5]; the novelty in that work was the presence of Dirichlet, Robin and Neumann boundary conditions.

So far, various properties of the DPME are classical; we mention a couple of them here. An interesting feature is that if the initial condition is non negative then the same is true for the solution at any time $t \geq 0$. Another characteristic is that solutions of DPME evolve with finite speed of propagation (in opposition to the heat equation, whose solutions spread out with infinite speed): see, for example, the Barenblatt solution as an example of the finite speed of propagation property. The FPME is less studied in the literature than its diffusive counterpart. Despite that, we refer the interested reader to [9, 10], where the authors studied properties of the weak solutions of the FPME.

Very recently in [6] the FPME (for $m = 2$) was obtained as the hydrodynamic limit of a particle system. Our goal in these notes is to generalize the results of [6] to any $m \in \mathbb{N}$, presenting the derivation of (1) as the hydrodynamic limit of a collection of SIPS whose jump rates are dependent on the number of particles close to where the jump occurs. In every model of this collection the exclusion rule is enforced, i.e., jumps to occupied sites are suppressed. With this dynamics, each site is occupied by at most one particle, therefore we are always dealing with exclusion process, highly studied in the SIPS literature. The case $m = 1$ (where the fractional heat equation is derived) was first presented in [14] and generalized in [8] in the presence of a slow barrier (where fractional Neumann and Robin boundary conditions were obtained). Since the linear setting was introduced a while ago in [14], we focus on the nonlinear one in this work.

Besides the exclusion rule, for $m \geq 2$ jumps from x to y are only allowed if there exists a set (among some possibilities, depending on x and y) of $m - 1$ sites at the neighborhood of the initial and final position which is fully occupied. We then show that the FPME can be obtained as the hydrodynamic limit of this SIPS. We do so by employing the tools of the classical entropy method developed in [13]. A direct consequence of this approach is the existence of weak solutions for the FPME, but we still need to prove the uniqueness of those solutions. We highlight that the same general strategy is applied to the case $m = 1$ (as it was done in [8, 14]). In these notes, by a matter of size we could not give all the technical details of the proofs of our results. Instead, we tried to choose the ones which are the most relevant and most enlightening for the case $m \geq 2$.

Here follows some open problems that arise naturally from our study; the first one is regarding the derivation of (1) when $m \notin \mathbb{N}$. Recently in [12] the answer to the previous question was provided for the diffusive case and for $m \in (0, 2)$; and the constructed model is a kind of reinforcement/penalization of the symmetric simple exclusion process. We believe that similar arguments could provide good answers in the fractional case and we leave this to future work. Another compelling question has to do with the extension of our results to higher dimensions, and also adding different boundary conditions such as Dirichlet, Robin or Neumann to the hydrodynamic equation. So far, fractional boundary conditions have been obtained in [2, 8], but these works only deal with the linear case. We are also interested in being able to apply the results we have to extend the list of fractional PDEs that one can obtain from the hydrodynamic limit of SIPS: either by choosing an exclusion process that allows more than a particle per site or by considering systems of SIPS in order to obtain systems of coupled FPMEs.

Outline of the article: In Sect. 2 we introduce the microscopic models that we analyse, we provide the notion of weak solutions for the fractional equations that we obtain and we state our main result, namely Theorem 9. In Sect. 3 we give an heuristic argument to illustrate how the notion of weak solution of our equations arises from the underlying particle system. In order to do so, several intermediate results are assumed :we refer the reader to [3, 6, 8] for the proofs. In Sect. 4 we prove our main theorem, i.e. the hydrodynamic limit. More precisely, in Sects. 4.1, 4.2 and 4.3 we obtain tightness, characterize the limit points and prove some energy estimates. Finally, in Sect. 4.4 we derive some results that are needed along the arguments in the previous subsections.

2 Models and Statement of Results

2.1 Microscopic Models: Exclusion and Porous Medium

In this section we introduce the two models that we will analyse in these notes. The first one is the well known symmetric exclusion process with long range interactions. The second is again an exclusion process, but the exchange rate depends strongly on the occupancy close to the sites where the exchanges occur. Both processes will be evolving in $\mathbb{Z}$, the set of integer numbers, whose elements are called *sites* and denoted by Latin letters such as x, y, z. Now we explain in detail the two dynamics. First we observe that our processes belong to the collection of exclusion processes which only allow at most one particle per site. Therefore, the state space of our models is the set $\Omega = \{0, 1\}^{\mathbb{Z}}$. The elements of this space are called configurations and we denote them by Greek letters such as η. Given a configuration $\eta \in \Omega$ and a site $x \in \mathbb{Z}$, we can only have two values for $\eta(x)$, i.e. $\eta(x) \in \{0, 1\}$ and we interpret that the site x is empty if $\eta(x) = 0$, and that the site x is occupied if $\eta(x) = 1$.

Particles jump in $\mathbb{Z}$ according to a probability measure $p : \mathbb{Z} \to [0, 1]$ defined by

$$\forall z \in \mathbb{Z}, \quad p(z) = \frac{c_\gamma}{|z|^{-\gamma-1}} \mathbb{1}_{\{z \neq 0\}}, \tag{2}$$

where $0 < \gamma < 2$ is fixed and c_γ is a normalizing constant that turns $p(\cdot)$ into a probability.

Given a configuration η and sites $x, y \in \mathbb{Z}$, we denote by $\eta^{x,y}$, the configuration obtained from η where the occupation of the sites x and y are exchanged, i.e.

$$\eta^{x,y}(z) := \begin{cases} \eta(z), & z \neq x, y, \\ \eta(y), & z = x, \\ \eta(x), & z = y. \end{cases}$$

To properly define the infinitesimal generators of our Markov processes we need to introduce the notion of local functions. Therefore, $f : \Omega \to \mathbb{R}$ is a *local function*, if there exists a finite $\Lambda \subset \mathbb{Z}$ such that $\forall x \in \Lambda, \eta_1(x) = \eta_2(x) \Rightarrow f(\eta_1) = f(\eta_2)$. With this definition in hands, for $m \in \mathbb{N}$ fixed, the continuous time Markov processes that we consider and that we denote by $(\eta_t)_{t \geq 0}$ have infinitesimal generator $\mathscr{L}$ given on local functions $f : \Omega \to \mathbb{R}$ by

$$\begin{aligned}(\mathscr{L} f)(\eta) :=& \frac{1}{4} \sum_{x,y} p(y-x) c_{x,y}^{(m)}(\eta) \xi_{x,y}(\eta) [f(\eta^{x,y}) - f(\eta)] \\ =& \frac{1}{4} \sum_{x,y} p(y-x) c_{x,y}^{(m)}(\eta) [f(\eta^{x,y}) - f(\eta)]. \end{aligned} \tag{3}$$

Above, $\xi_{x,y}(\eta) := \eta(x)[1-\eta(y)] + \eta(y)[1-\eta(x)] = [\eta(y)-\eta(y)]^2$, for any $x, y \in \mathbb{Z}$. The equality in (3) holds due to $\xi_{x,y}(\eta)[f(\eta^{x,y}) - f(\eta)] = f(\eta^{x,y}) - f(\eta)$. Above and hereinafter we will always assume that the discrete variables in a summation range over $\mathbb{Z}$, unless it is stated otherwise.

Before stating the general expression of $c^{(m)}_{x,y}(\eta)$, we give some motivation for its definition. For $m = 1$, in order to recover the dynamics described in [14] and in [8] (in the later for the choice $\alpha = 1$ and $\beta = 0$), we define $c^{(1)}_{x,y}(\eta) = 2$. In this work, $c^{(m)}_{x,y}(\eta)$ is always equals to twice the rate of a exchange of particles between x and y, according to the configuration η. For $m \geq 2$, in order to produce the diffusive porous medium equation $\partial \rho = \Delta \rho^m$, the role of $c^{(m)}_{x,y}(\eta)$ in (3) is fulfilled by

$$c^{(m,dif)}_{x,x+1}(\eta) := \sum_{k=1}^{m} \prod_{\substack{j=k-m \\ j \notin \{0,1\}}}^{k} \eta(x+j). \tag{4}$$

For details see [5]. In particular, we get

$$\begin{aligned}
c^{(2,dif)}_{x,x+1}(\eta) &= \eta(x-1) + \eta(x+2) = \eta(x-1) + \eta\big((x+1)+1\big);\\
c^{(3,dif)}_{x,x+1}(\eta) &= \eta(x-2)\eta(x-1) + \eta\big((x+1)+1\big)\big[\eta(x-1) + \eta\big((x+1)+2\big)\big];\\
c^{(4,dif)}_{x,x+1}(\eta) &= \eta(x-3)\eta(x-2)\eta(x-1) + \eta(x-2)\eta(x-1)\eta\big((x+1)+1\big)\\
&\quad + \eta\big((x+1)+1\big)\eta\big((x+1)+2\big)\big[\eta(x-1) + \eta\big((x+1)+3\big)\big].
\end{aligned}$$

In particular, replacing $x+1$ by y, we get

$$\begin{aligned}
c^{(2,dif)}_{x,y}(\eta) &= \eta(x-1) + \eta(y+1);\\
c^{(3,dif)}_{x,y}(\eta) &= \eta(x-2)\eta(x-1) + \eta(x-1)\eta(y+1) + \eta(y+1)\eta(y+2);\\
c^{(4,dif)}_{x,y}(\eta) &= \eta(x-3)\eta(x-2)\eta(x-1) + \eta(x-2)\eta(x-1)\eta(y+1)\\
&\quad + \eta(x-1)\eta(y+1)\eta(y+2) + \eta(y+1)\eta(y+2)\eta(y+3).
\end{aligned}$$

In order to have symmetric rates (i.e., $c^{(m)}_{x,y}(\eta) = c^{(m)}_{y,x}(\eta)$), in [6] we made the choice

$$c^{(2)}_{x,y}(\eta) := c^{(2,dif)}_{x,y}(\eta) + c^{(2,dif)}_{y,x}(\eta) = \eta(x-1) + \eta(y+1) + \eta(y-1) + \eta(x+1).$$

Note that in last case ($m = 2$) the jump rate can be multiplied by a factor $r \in \{0, 1, 2, 3, 4\}$ depending on the occupancy of the sites $\{x-1, x+1, y-1, y+1\}$, see the figure below (Fig. 1).

In [6], the following remark was crucial.

Remark 1 For the particular case where $y = x + 1$, we get

$$\xi_{x,y}(\eta)c^{(2)}_{x,y}(\eta) = \xi_{x,x+1}(\eta)[\eta(x-1) + \eta(x) + \eta(x+1) + \eta(x+2)]$$

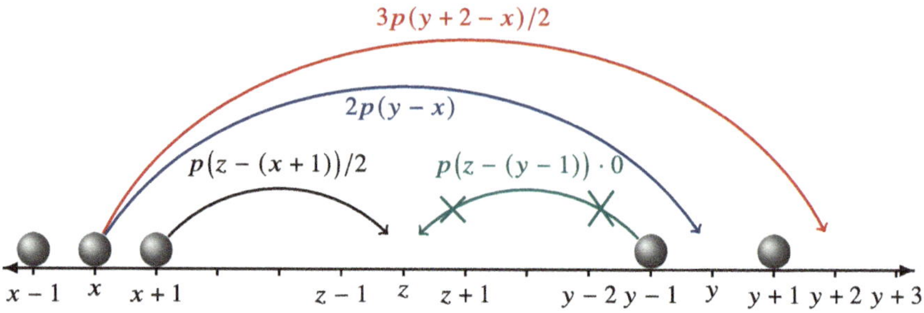

Fig. 1 Fractional porous medium model with long jumps. The rate of a jump between x_0 and y_0 is equal to half of the number of sites in $\{x_0 - 1, x_0 + 1, y_0 - 1, y_0 + 1\}$ which are occupied. For instance, the jump from x to y has rate $4/2$, since all the neighbors of x and y are occupied. On the other hand, the jump from $y - 1$ to z has rate zero (even z being an empty site), since neither of the sites z or $y - 1$ is next to an occupied site

$$=\xi_{x,x+1}(\eta)[\eta(x-1)+\eta(x+2)+1] \geq \xi_{x,x+1}(\eta).$$

Above we used the fact that $\eta(z) \in \{0, 1\}$ for every $z \in \mathbb{Z}$. In particular, the porous medium dynamics (for $m = 2$) always allow jumps of size 1, as long as the exclusion rule is enforced.

However, for every $m \geq 3$ it is *not true* that

$$\forall x \in \mathbb{Z}, \forall \eta \in \Omega, \quad \xi_{x,x+1}(\eta)[c^{(m,dif)}_{x,x+1}(\eta) + c^{(m,dif)}_{x+1,x}(\eta)] \geq \xi_{x,x+1}(\eta).$$

Therefore, in order to assure that jumps of size 1 are always allowed (which is fundamental to avoid blocked configurations), if $m \geq 3$ we add (a multiple of) the rates of the symmetric simple exclusion process in our dynamics. Then finally, we can give the general expression for $c^{(m)}_{x,y}(\eta)$.

$$c^{(m)}_{x,y}(\eta) := \begin{cases} 2, & m = 1; \\ c^{(m,dif)}_{x,y}(\eta) + c^{(m,dif)}_{y,x}(\eta) + \mathbb{1}_{\{m \geq 3\}} \mathbb{1}_{\{|x-y|=1\}}, & m \geq 2. \end{cases} \tag{5}$$

Remark 2 For every $m \geq 2$, it holds

$$c^{(m,dif)}_{x,y}(\eta) + c^{(m,dif)}_{y,x}(\eta) := \sum_{j=1}^{m} \Big(\prod_{i=0}^{m-2} \eta(a_{i+j,x,y}) + \prod_{i=0}^{m-2} \eta(a_{i+j+2(m-1),x,y}) \Big),$$

where for every $x, y \in \mathbb{Z}$, $\{a_{1,x,y}, a_{2,x,y}, \ldots a_{4m-4,x,y}\} \subset \mathbb{Z}$ is defined by

$$a_{j,x,y} := \begin{cases} x-(m-j), & j \in \{1,\dots,m-1\}; \\ y+j-(m-1), & j \in \{(m-1)+1,\dots,2(m-1)\}; \\ y-(3m-2-j), & j \in \{2(m-1)+1,\dots,3(m-1)\}; \\ x+j-3(m-1), & j \in \{3(m-1)+1,\dots,4(m-1)\}. \end{cases}$$

This means that if $|x-y|>1$, a particle can only jump between x and y if at least one of the $2m$ windows $\{a_{1,x,y},\dots,a_{m-1,x,y}\},\dots,\{a_{m,x,y},\dots,a_{2m-2,x,y}\}$, $\{a_{2m-1,x,y},\dots,a_{3m-3,x,y}\},\dots,\{a_{3m-2,x,y},\dots,a_{4m-4,x,y}\}$ is fully occupied.

An important feature of all these models is that their invariant measures are the Bernoulli product measures with parameter $b \in (0,1)$ denoted by ν_b and defined on Ω by their marginals given on $y \in \mathbb{Z}$ by $\nu_b\{\eta \in \Omega : \eta(y) = 1\} = b$. This means that under ν_b, the random variables $(\eta(y))_{y\in\mathbb{Z}}$ are independent and all have Bernoulli distribution of parameter b. Moreover, for every $\eta \in \Omega$ and for every $x, y \in \mathbb{Z}$ it holds

$$\forall x, y \in \mathbb{Z}, \forall \eta \in \Omega, \quad \nu_b(\eta^{x,y}) = \nu_b(\eta). \tag{6}$$

Thus, since $p(\cdot)$ given in (2) is symmetric, the measure ν_b is reversible and in particular, it is also invariant.

2.2 *Hydrodynamic Equations*

For $\gamma \in (0,2)$, the fractional Laplacian $-(-\Delta)^{\gamma/2}$ of exponent $\gamma/2$ is defined on the set of functions $G : \mathbb{R} \to \mathbb{R}$ such that $\int_{\mathbb{R}} \frac{G(u)}{(1+|u|)^{1+\gamma}}\,du < \infty$ by

$$[-(-\Delta)^{\gamma/2}G](u) := c_\gamma \lim_{\varepsilon\to 0^+} \int_{\mathbb{R}} \mathbb{1}_{\{|u-v|\geq\varepsilon\}} \frac{G(v)-G(u)}{|u-v|^{1+\gamma}}\,dv \tag{7}$$

provided the limit exists. Above, c_γ is the constant appearing in (2). An equivalent definition for the fractional Laplacian given in last display is through the Fourier transform, i.e. $-\widehat{(-\Delta)^{\gamma/2}G}(\xi) = |\xi|^\gamma \widehat{G}(\xi)$, nevertheless, we will not use this definition in what follows. Fix a measurable function $g : \mathbb{R} \to [0,1]$. In this article we are interested in analysing solutions of the Eq. (1). In that equation we assume that $m \in \mathbb{N}$. When $m = 1$, the Eq. (1) is the fractional heat equation, which is a linear equation; while for $2 \leq m \in \mathbb{N}$ it is the porous medium equation; which is a nonlinear equation. Now we define the space where our solutions belong.

2.2.1 Sobolev Spaces and Weak Solutions

Definition 3 The Sobolev space $\mathcal{H}^{\gamma/2}$ in $\mathbb{R}$ consists of all functions $f \in L^2(\mathbb{R})$ such that

$$[f]^2_{\mathcal{H}^{\gamma/2}} := \iint_{\mathbb{R}^2} \frac{[f(u) - f(v)]^2}{|u - v|^{1+\gamma}} \, du \, dv < \infty.$$

This is a Hilbert space for the norm $\| \cdot \|_{\mathcal{H}^{\gamma/2}}$ defined by

$$\| f \|^2_{\mathcal{H}^{\gamma/2}} := \int_{\mathbb{R}} [f(u)]^2 du + [f]^2_{\mathcal{H}^{\gamma/2}}.$$

Below we use the notation $\langle f, g \rangle$ to denote the inner product between two functions $f, g \in L^2(\mathbb{R})$; moreover, $N \subset L^2(\mathbb{R})$ is a normed vector space with norm $\| \cdot \|_N$.

Definition 4 We denote by $L^2(0, T; N)$ the set of all measurable functions $f : [0, T] \times \mathbb{R} \to \mathbb{R}$ such that $f(s, \cdot) \in N$ for almost every s on $[0, T]$ and $\int_0^T \| f(s, \cdot) \|^2_N \, ds < \infty$. Moreover, the set $P([0, T], N)$ is the space of functions $G : [0, T] \times \mathbb{R} \to \mathbb{R}$ such that there exist $k \in \{0, 1, 2, \ldots, \}$ and $G_0, G_1, \ldots, G_k \in N$ so that for all $(t, u) \in [0, T] \times \mathbb{R}$, $G(t, u) = \sum_{j=0}^k t^j G_j(u)$.

Given $r \in \{1, 2, \ldots\}$, $G : \mathbb{R} \to \mathbb{R}$ is in $C^r(\mathbb{R})$ if G is r times continuously differentiable and for $r = 0$, $C^0(\mathbb{R})$ denotes the set of continuous functions in $\mathbb{R}$. Also, $G \in C^r_c(\mathbb{R})$ if $G \in C^r(\mathbb{R})$ and G has compact support. Moreover, we use the notation $C^\infty_c(\mathbb{R}) := \cap_{r=0}^\infty C^r_c(\mathbb{R})$. Our space of test functions is $\mathcal{S} := P\big([0, T], C^\infty_c(\mathbb{R})\big)$. In what follows, given $H : [0, T] \times \mathbb{R} \to \mathbb{R}$, we denote $H(s, \cdot)$ by H_s, where $s \in [0, T]$.

Definition 5 (*Weak solutions*) Let $g : \mathbb{R} \to [0, 1]$ be a measurable function. We say that a function $\rho : [0, T] \times \mathbb{R} \to [0, 1]$ is a weak solution of the fractional porous medium Eq. (1) with initial condition g if the following conditions hold:

1. for every $t \in [0, T]$ and for every $G \in \mathcal{S}$, it holds $F(t, \rho, G, g) = 0$, where

$$\begin{aligned} F(t, \rho, G, g) :=& \langle \rho_t, G_t \rangle - \langle g, G_0 \rangle \\ & - \int_0^t \langle \rho_s, \partial_s G_s \rangle \, ds - \int_0^t \langle \rho_s^m, [-(-\Delta)^{\gamma/2} G_s] \rangle \, ds; \end{aligned} \tag{8}$$

2. there exists some $b \in (0, 1)$ satisfying both $\rho - b \in L^2\big(0, T; L^2(\mathbb{R})\big)$ and also $\rho^m - b^m \in L^2(0, T; \mathcal{H}^{\gamma/2})$.

Remark 6 In the particular case $m = 1$, the condition $\rho - b \in L^2\big(0, T; L^2(\mathbb{R})\big)$ is a direct consequence of $\rho^m - b^m \in L^2(0, T; \mathcal{H}^{\gamma/2})$.

Remark 7 The uniqueness of weak solutions as given in the previous definition is proved in Appendix C of [6].

2.3 Hydrodynamic Limit

Let us now fix a time $T > 0$ and a finite time horizon $[0, T]$. Since we want to observe a non-trivial evolution, we consider our Markov processes speeded up in the time scale n^γ for $\gamma \in (0, 2)$. We will explain later the reason for the choice of the time scale defined above. We use the notation $\eta_t^n := \eta_{tn^\gamma}$ for the speeded up process and note that it has the generator $\mathcal{L}_n := n^\gamma \mathcal{L}$.

Since in our models the unique conserved quantity is the density of particles, we define the empirical measure as the measure that gives weight $1/n$ to each particle in the following way:

$$\pi^n(\eta, du) := \frac{1}{n} \sum_x \eta(x) \delta_{x/n}(du),$$

where δ_u is a Dirac mass on $u \in \mathbb{R}$. Now we define the process of the empirical measures as $\pi_t^n(\eta, du) := \pi^n(\eta_t^n, du)$. For a measurable function $G : \mathbb{R} \to \mathbb{R}$, we denote the integral of G with respect to the empirical measure π_t^n, by $\langle \pi_t^n, G \rangle$.

Now we define the conditions on the initial measures. To that end, for every $n \geq 1$, let μ_n be a probability measure on our state space Ω. Let $\mathcal{M}^+$ be the space of non-negative Radon measures on $\mathbb{R}$ and equipped with the weak topology. Let $\mathbb{P}_{\mu_n}$ be the probability measure on the Skorokhod space $\mathcal{D}([0, T], \Omega)$ induced by the Markov process $(\eta_t)_{t \in [0,T]}$ and the initial measure μ_n. Moreover, let $\mathbb{Q}_n$ be the probability measure on $\mathcal{D}([0, T], \mathcal{M}^+)$ induced by $(\pi_t^n)_{t \in [0,T]}$ and $\mathbb{P}_{\mu_n}$.

Definition 8 Let $g : \mathbb{R} \to [0, 1]$ be a measurable function and $(\mu_n)_{n \geq 1}$ a sequence of probability measures in Ω. We say that $(\mu_n)_{n \geq 1}$ is *associated with* $g(\cdot)$ if for any $G \in C_c^0(\mathbb{R})$ and any $\delta > 0$,

$$\lim_{n \to \infty} \mu_n\Big(\eta \in \Omega : \Big|\frac{1}{n} \sum_x G\big(\tfrac{x}{n}\big)\eta(x) - \int_{\mathbb{R}} G(u) g(u)\, du \Big| > \delta\Big) = 0.$$

The previous assumption requires the validity of a law of large numbers for the empirical measure at time $t = 0$., i.e. the sequence of random measures $\{\pi_0^n\}_{n \in \mathbb{N}}$ converges, as $n \to \infty$ to the deterministic measure $\pi_0(du) := g(u)du$. Under last assumption, our goal is to show that for any time t the same result is true but the density will be a weak solution to a PDE, called the *hydrodynamic equation.*

In some of our results below we will need some extra conditions on the initial measure of the systems. To that end we state the following condition, that will be imposed for some models:

$$\textbf{Assumption (A)}: \;\; \exists C_b > 0 : \; \forall n \geq 1, \quad H(\mu_n | \nu_b) \leq C_b n. \tag{9}$$

Now we are in position to state the main result of these notes.

Theorem 9 *Let $g : \mathbb{R} \to [0, 1]$ be a measurable function. Let $(\mu_n)_{n\geq 1}$ be a sequence of probability measures in Ω associated to the profile $g(\cdot)$ and satisfying Assumption (A) for some $b \in (0, 1)$. Then, for any $t \in [0, T]$, any $G \in C_c^0(\mathbb{R})$ and any $\delta > 0$,*

$$\lim_{n\to\infty} \mathbb{P}_{\mu_n}\Big(\eta_\cdot^n \in \mathscr{D}([0, T], \Omega) : \Big|\frac{1}{n}\sum_x G\big(\tfrac{x}{n}\big)\eta_t^n(x) - \int_{\mathbb{R}} G(u)\rho(t, u)\, du\Big| > \delta\Big) = 0,$$

where $\rho(t, \cdot)$ is the unique weak solution of (1).

Outline of the proof: For all the models we follow the entropy method introduced in [13]. To make the presentation as simple as possible we hide many technical results and we just focus on the most important steps of the proof. We refer the reader to [3, 6, 8] (and references there in) for more details. In Sect. 3 we give an heuristic argument, highlighting the main steps of the proof so that we can derive the integral notions of the weak solutions of our hydrodynamic equations. In Sect. 4.1, we prove that the sequence $(\mathbb{Q}_n)_{n\geq 1}$ is tight with respect to the Skorokhod topology of $\mathscr{D}([0, T], \mathscr{M}^+)$. Then, as a consequence of Prohorov's Theorem (see Theorem 6.1 in [4]), we are able to conclude that this sequence is relatively compact. From this we know that the sequence $(\mathbb{Q}_n)_{n\geq 1}$ has a convergent subsequence, i.e., there exists a subsequence $(\mathbb{Q}_{n_j})_{j\geq 1}$ and a measure $\mathbb{Q}$ such that the sequence $(\mathbb{Q}_{n_j})_{j\geq 1}$ weakly converges to $\mathbb{Q}$. In Sect. 4.2 (resp. Sect. 4.3) we prove that any such limit point $\mathbb{Q}$ is concentrated on trajectories of measures satisfying the first (resp. the second) condition of weak solutions of (1). Combining those results with the uniqueness of weak solutions of (1), we can conclude that the aforementioned limit point $\mathbb{Q}$ is actually unique, leading to the conclusion of Theorem 9. Finally in Sect. 4.4 we prove a lemma which is required in Sects. 4.2 and 4.3.

3 Heuristic Arguments for the Hydrodynamic Equations

In this section we present an heuristic reasoning which leads us to the integral equation (8) in Definition 5. We begin by assuming that the sequence $(\mathbb{Q}_n)_{n\geq 1}$ is tight (this fact will be proved in the next section) and we denote by $\mathbb{Q}$ a limit point. A simple computation based on the fact that our variables are bounded allows showing that the limit measure $\mathbb{Q}$ is concentrated on trajectories of measures $\pi_t(du)$ that are absolutely continuous with respect to the Lebesgue measure, i.e. $\pi_t(du) := \rho(t, u)du$. Now we need to characterize $\rho(t, u)$ as a weak solution to the corresponding fractional equation. According to Dynkin's formula (see Lemma $A.1.5.1$ of [15]),

$$\mathscr{M}_t^n(G) := \langle \pi_t^n, G_t\rangle - \langle \pi_0^n, G_0\rangle - \int_0^t \partial_s\langle \pi_s^n, G_s\rangle\, ds - \int_0^t n^\gamma \mathscr{L}\langle \pi_s^n, G_s\rangle\, ds \quad (10)$$

is a martingale with respect to $\mathscr{F}_t^n := \{\sigma(\eta_s^n) : s \leq t\}$, for every $n \geq 1$, $t \in [0, T]$ and $G \in \mathcal{S}$. Since the sequence $(\mathbb{Q}_n)_{n\geq 1}$ is tight, let n_j be a subsequence such that

$(\mathbb{Q}_{n_j})_{j\geq 1}$ weakly converges to $\mathbb{Q}$, which is supported on trajectories of the form $\pi_t(du) = \rho(t,u)du$. To make notation simple we assume that $n_j = n$. From this it follows that the first three terms on the right-hand side of (10) converge, as $n \to \infty$, in $L^1(\mathbb{P}_{\mu_n})$ to

$$\int_{\mathbb{R}} \rho_t(u) G_t(u)\, du - \int_{\mathbb{R}} \rho_0(u) G_0(u)\, du - \int_0^t \int_{\mathbb{R}} \rho_s(u) \partial_s G_s(u)\, du\, ds.$$

From Definition 8, we get that for any $\delta > 0$,

$$\mathbb{Q}\Big(\Big| \int_{\mathbb{R}} [\rho_0(u) - g(u)] G_0(u)\, du \Big| > \delta\Big) = 0.$$

Hence last display converges in $L^1(\mathbb{P}_{\mu_n})$ to

$$\int_{\mathbb{R}} \rho_t(u) G_t(u)\, du - \int_{\mathbb{R}} g(u) G_0(u)\, du - \int_0^t \int_{\mathbb{R}} \rho_s(u) \partial_s G_s(u)\, du\, ds,$$

as $n \to \infty$. Now we focus on last term of (10), which is known in the literature as the *integral term*; it describes the action of the infinitesimal generator in the empirical measure associated to the conserved quantity: the density of particles.

The integral term will lead us to the fractional heat equation (resp. fractional porous medium equation) for $m = 1$ (resp. $2 \leq m \in \mathbb{N}$). Indeed, by fixing $x_0 \in \mathbb{Z}$ and making $f = \eta(x_0)$ in (3), we have

$$\mathcal{L}\big(\eta(x_0)\big) = \frac{1}{2} \sum_y p(y - x_0) c^{(m)}_{x_0,y}(\eta)[\eta(y) - \eta(x_0)]. \tag{11}$$

Above we used the facts that $[\eta^{y,z}(x_0) - \eta(x_0)] = 0$ when $y, z \in \mathbb{Z} - \{x_0\}$ and that $p(y - x_0) c^n_{x_0,y}(\eta) = p(x_0 - y) c^n_{y,x_0}(\eta)$. From the linearity of $\mathcal{L}$ and (11), we get

$$n^\gamma \mathcal{L}(\langle \pi^n_s, G_s\rangle) = \frac{n^\gamma}{2n} \sum_{x,y} G_s(\tfrac{x}{n}) p(y - x) c^{(m)}_{x,y}(\eta^n_s)[\eta^n_s(y) - \eta^n_s(x)]. \tag{12}$$

Next we analyse (12) for particular values of m.

3.1 Heuristics for m = 1

For $m = 1$, we get from (5) that $c^{(m)}_{x,y}(\eta^n_s)[\eta^n_s(y) - \eta^n_s(x)] = 2[\eta^n_s(y) - \eta^n_s(x)]$, therefore the right-hand side of (12) can be rewritten as

$$\frac{n^\gamma}{n}\sum_{x,y} G_s(\tfrac{x}{n})p(y-x)\eta_s^n(y) - \frac{n^\gamma}{n}\sum_{x,y} G_s(\tfrac{x}{n})p(y-x)\eta_s^n(x)$$
$$=\frac{1}{n}\sum_{x,y} n^\gamma [G_s(\tfrac{y}{n}) - G_s(\tfrac{x}{n})]p(y-x)\eta_s^n(x).$$

In last equality we used the symmetry of $p(\cdot)$. Therefore,

$$\int_0^t n^\gamma \mathcal{L}\langle \pi_s^n, G_s\rangle\, ds = \int_0^t \frac{1}{n}\sum_x n^\gamma \mathcal{K}_n G_s(\tfrac{x}{n})\eta_s^n(x)\, ds, \tag{13}$$

where $\mathcal{K}_n$ is defined on functions $G \in \mathcal{S}$ by

$$\mathcal{K}_n G_s(\tfrac{x}{n}) := \sum_y \left[G_s(\tfrac{y}{n}) - G_s(\tfrac{x}{n})\right] p(y-x). \tag{14}$$

The term on the right-hand side of (13) can be treated with next result, which motivates the choice n^γ for the choice of the time scale. We refer the reader to Proposition A.1 of [8] for its proof.

Proposition 10 *For every $\gamma \in (0, 2)$ and $G \in \mathcal{S}$, it holds*

$$\lim_{n\to\infty} \frac{1}{n}\sum_x \sup_{s\in[0,T]} \left|n^\gamma \mathcal{K}_n G_s\left(\tfrac{x}{n}\right) - [-(-\Delta)^{\gamma/2} G_s]\left(\tfrac{x}{n}\right)\right| = 0.$$

Note that (13) can be written as $\int_0^t \langle \pi_s^n, n^\gamma \mathcal{K}_n G_s\rangle$ and so it is written as a function of the empirical measure. Thanks to Proposition 10, we conclude that for $m = 1$, the integral term converges, as $n \to \infty$, in $L^1(\mathbb{P}_{\mu_n})$ to

$$\int_0^t \int_{\mathbb{R}} \rho_s(u)[-(-\Delta)^{\gamma/2} G_s](u)\, du\, ds,$$

and when we collect all terms we get the integral equation in Definition 5.

3.2 Heuristics for $m \geq 2$

For $m \geq 2$, we will perform some algebraic manipulations in order to rewrite the integral term as the sum of two terms. The first one will be the *principal term*, which will converge as $n \to \infty$ and in $L^1(\mathbb{P}_{\mu_n})$ to

$$\int_0^t \int_{\mathbb{R}} \rho_s^m(u)[-(-\Delta)^{\gamma/2} G_s](u)\, du\, ds, \tag{15}$$

which is the final term to be obtained in the integral equation in Definition 5. The second one is the *extra term*, that vanishes as $n \to \infty$ and in $L^1(\mathbb{P}_{\mu_n})$.

3.2.1 Heuristics for $m = 2$

We begin by treating the particular case $m = 2$, in the same way as it was done in [6]. In this case, Remark 2 leads to

$$
\begin{aligned}
&[c_{x,y}^{(m,dif)}(\eta) + c_{y,x}^{(m,dif)}(\eta)][\eta(y) - \eta(x)]\\
=&[\eta(x-1) + \eta(y+1) + \eta(y-1) + \eta(x+1)][\eta(y) - \eta(x)]\\
=&[\eta(y)\eta(y-1) + \eta(y)\eta(y+1)] - [\eta(x)\eta(x-1) + \eta(x)\eta(x+1)]\\
+&[\eta(y)\eta(x-1) - \eta(y+1)\eta(x)] - [\eta(x)\eta(y-1) - \eta(x+1)\eta(y)].
\end{aligned}
$$

In the third line of last display, we simply collected the products of η's that either depend only on x or depend only on y. And in last line, we collected all the remaining products (which depend both on x and y). The principal term (resp. extra term) will be produced by the terms in the third line (resp. the last line). The extra term vanishes in $L^1(\mathbb{P}_{\mu_n})$ as $n \to \infty$, due to Proposition 11 below. Indeed, combining last display with an exchange of variables, the right-hand side of (12) can be rewritten as

$$
\begin{aligned}
&\frac{1}{2n}\sum_{x,y} n^\gamma [G_s(\tfrac{y}{n})p(x-y) - G_s(\tfrac{x}{n})p(y-x)][\eta_s^n(x)\eta_s^n(x-1) + \eta_s^n(x)\eta_s^n(x+1)]\\
+&\frac{1}{2n}\sum_{x,y} n^\gamma [G_s(\tfrac{y}{n})p(x-y) - G_s(\tfrac{x}{n})p(y-x)][\eta_s^n(x)\eta_s^n(y-1) - \eta_s^n(x+1)\eta_s^n(y)].
\end{aligned}
$$

From the symmetry of $p(\cdot)$, last display can be rewritten as

$$
\begin{aligned}
&\frac{1}{2n}\sum_{x} n^\gamma \mathcal{K}_n G_s(\tfrac{x}{n})[\eta_s^n(x)\eta_s^n(x-1) + \eta_s^n(x)\eta_s^n(x+1)]\\
+&\frac{1}{2n}\sum_{x,y} n^\gamma [G_s(\tfrac{y}{n}) - G_s(\tfrac{x}{n})]p(y-x)[\eta_s^n(x)\eta_s^n(y-1) - \eta_s^n(x+1)\eta_s^n(y)].
\end{aligned}
$$

Above, the term in first line (resp. second line) is the principal term (resp. extra term). Contrarily to what we have seen in (13), now the principal term cannot be straightforwardly written in terms of the empirical measure (and this is the case for all $m \geq 2$ and for this reason extra arguments are needed in order to write this term as a function of the empirical measure).

3.2.2 Heuristics for $m = 3$

For $m = 3$, we have an extra contribution of the SSEP dynamics that we added in order to have always the possibility of performing jumps of length one (in the previous case this was granted by the porous medium dynamics). In this case, from Remark 2, we have

$$
\begin{aligned}
&[c_{x,y}^{(m,dif)}(\eta) + c_{y,x}^{(m,dif)}(\eta)][\eta(y) - \eta(x)] \\
=&[\eta(y)\eta(y-1)\eta(y-2) + \eta(y)\eta(y+1)\eta(y+2)] \\
-&[\eta(x)\eta(x-1)\eta(x-2) + \eta(x)\eta(x+1)\eta(x+2)] \\
+&\big[\eta(y)\eta(x-1)\eta(x-2) + \eta(y)\eta(y+1)\eta(x-1) \\
-&\eta(y+1)\eta(x)\eta(x-1) - \eta(y+1)\eta(y+2)\eta(x)\big] \\
-&\big[\eta(x)\eta(y-1)\eta(y-2) + \eta(x)\eta(x+1)\eta(y-1) \\
-&\eta(x+1)\eta(y)\eta(y-1) - \eta(x+1)\eta(x+2)\eta(y)\big].
\end{aligned}
$$

In the second and third lines of last display, we simply collected the products of η's which either depend only on x or depend only on y. On the other hand, in the last four lines, we collected all the remaining products (which depend of both x and y). The principal term (resp. extra term) will be produced by the terms in the second and third lines (resp. last four lines). The extra term converges in $L^1(\mathbb{P}_{\mu_n})$ to zero as $n \to \infty$, due to Proposition 11 below. Indeed, combining last display with an exchange of variables, the right-hand side of (12) can be rewritten as

$$
\begin{aligned}
&\int_0^t \frac{n^\gamma}{2n} \sum_{x,y} [G_s(\tfrac{y}{n})p(x-y) - G_s(\tfrac{x}{n})p(y-x)] \times \\
&\times [\eta_s^n(x)\eta_s^n(x-1)\eta_s^n(x-2) + \eta_s^n(x)\eta_s^n(x+1)\eta_s^n(x+2)]\,ds \\
+&\int_0^t \frac{n^\gamma}{2n} \sum_{x,y} [G_s(\tfrac{y}{n})p(x-y) - G_s(\tfrac{x}{n})p(y-x)] \times \\
&\times \big(\eta_s^n(x)\eta_s^n(y-1)[\eta_s^n(y-2) + \eta_s^n(x+1)] - \eta_s^n(x+1)\eta_s^n(y)[\eta_s^n(y-1) + \eta_s^n(x+2)]\big)\,ds \\
+&\int_0^t \frac{n^\gamma}{2n} \sum_{x} \big(G_s(\tfrac{x+1}{n})p(-1) + G_s(\tfrac{x-1}{n})p(1) - G_s(\tfrac{x}{n})[p(1) + p(-1)]\big)\eta_s^n(x)\,ds.
\end{aligned}
$$

From the symmetry of $p(\cdot)$, last display can be rewritten as

$$
\begin{aligned}
&\int_0^t \frac{1}{2n} \sum_{x} n^\gamma \mathcal{K}_n G_s(\tfrac{x}{n})[\eta_s^n(x)\eta_s^n(x-1)\eta_s^n(x-2) + \eta_s^n(x)\eta_s^n(x+1)\eta_s^n(x+2)]\,ds \\
+&\int_0^t \Big\{\frac{1}{2n} \sum_{x,y} n^\gamma [G_s(\tfrac{y}{n}) - G_s(\tfrac{x}{n})]p(y-x)\eta_s^n(x)\eta_s^n(y-1)[\eta_s^n(y-2) + \eta_s^n(x+1)] \\
-&\frac{1}{2n} \sum_{x,y} n^\gamma [G_s(\tfrac{y}{n}) - G_s(\tfrac{x}{n})]p(y-x)\eta_s^n(x+1)\eta_s^n(y)[\eta_s^n(y-1) + \eta_s^n(x+2)]\Big\}\,ds \\
+&\int_0^t \frac{p(1)}{2n} \sum_{x} n^\gamma [G_s(\tfrac{x+1}{n}) + G_s(\tfrac{x-1}{n}) - 2G_s(\tfrac{x}{n})]\eta_s^n(x)\,ds.
\end{aligned}
$$

The term in the fourth line of last display comes from the contribution of the SSEP and it vanishes as $n \to \infty$, since it is not scaled on the diffusive scale n^2 but on n^γ.

3.2.3 Heuristics for the General Case $m \geq 2$

For every $m \geq 2$, from Remark 2, it holds

$$
\begin{aligned}
&[c_{x,y}^{(m,dif)}(\eta) + c_{y,x}^{(m,dif)}(\eta)][\eta(y) - \eta(x)] = B_m(\eta, y) - B_m(\eta, x) \\
&+[C_m(\eta, y, x) - C_m(\eta, y+1, x+1)] - [C_m(\eta, x, y) - C_m(\eta, x+1, y+1)],
\end{aligned} \tag{16}
$$

where $B_m : \Omega \times \mathbb{Z} \to [0, 2]$ and $C_m : \Omega \times \mathbb{Z} \times \mathbb{Z} \to [0, m-1]$ are given by

$$
B_m(\eta, z) := \prod_{i=0}^{m-1} \eta(z - i) + \prod_{i=0}^{m-1} \eta(z + i); \tag{17}
$$

$$
C_m(\eta, z, w) := \sum_{k=1}^{m-1} \prod_{j=0}^{k-1} \eta(z + j) \times \prod_{i=1}^{m-k} \eta(w - i). \tag{18}
$$

Combining (16) with an exchange of variables, the right-hand side of (12) can be rewritten as

$$
\begin{aligned}
&\frac{n^\gamma}{2n} \sum_{x,y} [G_s(\tfrac{y}{n}) p(x - y) - G_s(\tfrac{x}{n}) p(y - x)] B_m(\eta_s^n, x) \\
&+\frac{n^\gamma}{2n} \sum_{x,y} [G_s(\tfrac{y}{n}) p(x - y) - G_s(\tfrac{x}{n}) p(y - x)][C_m(\eta_s^n, x, y) - C_m(\eta_s^n, x+1, y+1)] \\
&+\mathbb{1}_{\{m \geq 3\}} \frac{n^\gamma}{2n} \sum_{x} \big([G_s(\tfrac{x+1}{n}) p(-1) - G_s(\tfrac{x}{n}) p(1)] + [G_s(\tfrac{x-1}{n}) p(1) - G_s(\tfrac{x}{n}) p(-1)]\big) \eta_s^n(x).
\end{aligned}
$$

For every $G \in \mathcal{S}$, define ∇G_s^n by $\nabla G_s^n(\frac{x}{n}) = n[G_s(\frac{x}{n}) - G_s(\frac{x-1}{n})]$. Then, from the symmetry of $p(\cdot)$, the integral term can be rewritten as

$$
\int_0^t \frac{1}{2n} \sum_{x} n^\gamma \mathcal{K}_n G_s(\tfrac{x}{n}) B_m(\eta_s^n, x) \, ds \tag{19}
$$

$$
+ \int_0^t \frac{1}{2n^2} \sum_{x,y} n^\gamma [\nabla G_s^n(\tfrac{y}{n}) - \nabla G_s^n(\tfrac{x}{n})] p(y - x) C_m(\eta_s^n, x, y) \, ds \tag{20}
$$

$$
+\mathbb{1}_{\{m \geq 3\}} \int_0^t \frac{p(1)}{2n} \sum_{x} n^\gamma [G_s(\tfrac{x+1}{n}) + G_s(\tfrac{x-1}{n}) - 2G_s(\tfrac{x}{n})] \eta_s^n(x) \, ds. \tag{21}
$$

In order to obtain (20), we performed the change of variables $(z, w) = (x+1, y+1)$. Thus, the principal term (resp. the extra term) is given by (19) (resp. the sum of (20) and (21)). In particular, the extra term can be rewritten as $\int_0^t \mathcal{R}_n^G(s)ds$, where

$$\begin{aligned}\mathcal{R}_n^G(s) :=& \frac{1}{2n^2}\sum_{x,y} n^\gamma[\nabla G_s^n(\tfrac{y}{n}) - \nabla G_s^n(\tfrac{x}{n})]p(y-x)C_m(\eta_s^n, x, y)\\ &+\mathbb{1}_{\{m\geq 3\}}\frac{p(1)}{2n}\sum_x n^\gamma[G_s(\tfrac{x+1}{n}) + G_s(\tfrac{x-1}{n}) - 2G_s(\tfrac{x}{n})]\eta_s^n(x).\end{aligned} \tag{22}$$

Combining (18) with the fact that the variables are bounded (i.e. $|\eta_s^n(\cdot)| \leq 1$), we get $|C_m(\cdot,\cdot,\cdot)| \leq m-1$. Thus, we have that $\sup_{s\in[0,T]}|\mathcal{R}_n^G(s)| \leq Y_1^{G,n} + Y_2^{G,n}$ for every $G \in \mathcal{S}$, where Y_1^G and Y_2^G are given by

$$\begin{aligned}Y_1^{G,n} :=& \frac{m-1}{n^2}\sum_{x,y}\sup_{s\in[0,T]} n^\gamma|\nabla G_s^n(\tfrac{y}{n}) - \nabla G_s^n(\tfrac{x}{n})|p(y-x);\\ Y_2^{G,n} :=& \frac{1}{n}\sum_x \sup_{s\in[0,T]} n^\gamma\left|G_s(\tfrac{x+1}{n}) + G_s(\tfrac{x-1}{n}) - 2G_s(\tfrac{x}{n})\right|.\end{aligned}$$

Next we claim that for all $G \in \mathcal{S}$, it holds $\lim_{n\to\infty} Y_1^{G,n} + Y_2^{G,n} = 0$. In order to treat $Y_1^{G,n}$ we apply Proposition 11 below. We refer the interested reader in its proof to Proposition 3.1 of [6].

Proposition 11 *For* $\gamma \in (0,2)$, *let* $\delta_\gamma = \frac{1}{2}\mathbf{1}_{\gamma=1} + \mathbf{1}_{\gamma\in(1,2)}$. *Then for any* $G \in \mathcal{S}$

$$Y_1^{G,n} \lesssim \max\left\{n^{\gamma-2}, n^{-1}, n^{\gamma-1-\delta_\gamma}\right\}.$$

In particular, since $\delta_\gamma > \gamma - 1$ *and* $\gamma < 2$, *last display vanishes as* $n \to \infty$.

In order to treat $Y_2^{G,n}$, we apply a second-order Taylor expansion of G_s around x/n to conclude that

$$\lim_{n\to\infty}\frac{1}{n}\sum_x \sup_{s\in[0,T]}\left|n^2[G_s(\tfrac{x+1}{n}) + G_s(\tfrac{x-1}{n}) - 2G_s(\tfrac{x}{n})] - \Delta G_s(\tfrac{x}{n})\right| = 0.$$

Combining this with the fact that $\gamma < 2$, we have that $\lim_{n\to\infty} Y_2^{G,n} = 0$ for every $G \in \mathcal{S}$ and this proves the claim. In particular, it holds

$$\lim_{n\to\infty}\mathbb{E}_{\mu_n}\Big[\sup_{t\in[0,T]}\Big|\int_0^t \mathcal{R}_n^G(s)\,ds\Big|\Big] = 0. \tag{23}$$

Finally, we treat the display in (19); from (17), it can be rewritten as

$$\int_0^t \frac{1}{2n}\sum_x n^\gamma\big[\mathcal{K}_n G_s(\tfrac{x}{n}) + \mathcal{K}_n G_s(\tfrac{x+m-1}{n})\big]\prod_{i=0}^{m-1}\eta_s^n(x+i)\,ds.$$

Due to the products of η_s^n's, it is not straightforward to close the integral term in terms of the empirical measure; in order to do so, we apply Lemma 12 below. Before stating this lemma, we introduce a bit of notation. For every $\ell \geq 1$ in $\mathbb{Z}$ and $x \in \mathbb{Z}$, we denote the empirical average in a box of side ℓ *at the right* of x by

$$\overrightarrow{\eta}^{\ell}(x) := \frac{1}{\ell}\sum_{i=1}^{\ell}\eta(x+i). \tag{24}$$

Lemma 12 (Replacement Lemma) *Let* $(\Phi_n)_{n\geq 1} : [0,T]\times\mathbb{R}\to\mathbb{R}$ *be a sequence of functions satisfying*

$$\frac{1}{n}\sum_x \sup_{s\in[0,T]}|\Phi_n(s,\tfrac{x}{n})| \leq M_1 \quad \textit{and} \quad \|\Phi_n\|_\infty := \sup_{(s,u)\in[0,T]\times\mathbb{R}}|\Phi_n(s,u)| \leq M_2, \tag{25}$$

for all $n \geq 1$ *and for some positive constants* M_1, M_2. *Under Assumption (A), for any* $m \geq 2$ *and every* $t \in [0,T]$, *it holds*

$$\varlimsup_{\varepsilon\to 0^+}\varlimsup_{n\to\infty}\mathbb{E}_{\mu_n}\Big[\Big|\int_0^t \frac{1}{n}\sum_x \Phi_n(s,\tfrac{x}{n})\Big\{\prod_{i=0}^{m-1}\eta_s^n(x+i) - \prod_{i=0}^{m-1}\overrightarrow{\eta}_s^{\varepsilon n}(x+i\varepsilon n)\Big\}ds\Big|\Big] = 0.$$

In order to apply Lemma 12, we need to ensure that (25) is satisfied. This is done by applying a classical result regarding the fractional Laplacian, Proposition 13 below. We refer the interested reader in its proof to Proposition 3.1 of [6] for more details.

Proposition 13 *For every* $G \in \mathcal{S}$, *we have*

$$\sup_{n\geq 1}\frac{1}{n}\sum_x \sup_{s\in[0,T]}\big|[-(-\Delta)^{\gamma/2}G_s](\tfrac{x}{n})\big| < \infty. \tag{26}$$

Thus, by combining Propositions 10 and 13 with Lemma 12, we have that (19) converges, as $n \to \infty$, to (15), in $L^1(\mathbb{P}_{\mu_n})$, and we get the integral equation in Definition 5.

4 Proof of Theorem 9

4.1 Tightness

Our goal is to prove that $(\mathbb{Q}_n)_{n\geq 1}$ is tight with respect to the Skorokhod topology of $\mathcal{D}([0,T],\mathcal{M}^+)$. Following Propositions 4.1.6 and 4.1.7 of [15], it is enough to show that

$$\lim_{\omega\to 0}\varlimsup_{n\to\infty}\sup_{\tau\in\mathcal{T}_T,\tilde{t}\leq\omega}\mathbb{P}_{\mu_n}\Big(\eta^n_\cdot\in\mathcal{D}([0,T],\Omega):\big|\langle\pi^n_{\tau+\tilde{t}},G\rangle-\langle\pi^n_\tau,G\rangle\big|>\varepsilon\Big)=0, \tag{27}$$

for every $G\in\mathcal{S}$ and every $\varepsilon>0$. Above, $\mathcal{T}_T$ is the set of stopping times bounded by T, therefore $\tau+\tilde{t}$ must be read as $\min\{\tau+\tilde{t},T\}$. From (10), Markov's and Chebyshev's inequalities, (27) is bounded from above by

$$\lim_{\omega\to 0}\varlimsup_{n\to\infty}\sup_{\tau\in\mathcal{T}_T,\tilde{t}\leq\omega}\frac{2}{\varepsilon}\mathbb{E}_{\mu_n}\Big[\Big|\int_\tau^{\tau+\tilde{t}}n^\gamma\mathcal{L}\langle\pi^n_r,G_r\rangle\,dr\Big|\Big] \tag{28}$$

$$+\lim_{\omega\to 0}\varlimsup_{n\to\infty}\sup_{\tau\in\mathcal{T}_T,\tilde{t}\leq\omega}\frac{4}{\varepsilon^2}\mathbb{E}_{\mu_n}\Big[|\mathcal{M}^n_{\tau+\tilde{t}}(G)-\mathcal{M}^n_\tau(G)|^2\Big]. \tag{29}$$

It is enough to prove that both limits in last display vanish as $n\to\infty$, for every $G\in\mathcal{S}$ fixed. We begin with the former one. Combining (19) and (22) with Proposition 10 and the fact that $\lim_{n\to\infty}Y_1^{G,n}+Y_2^{G,n}=0$, we can conclude, in particular, that for every $G\in\mathcal{S}$, there exists $C(G)$ such that $\sup_{s\in[0,T]}|n^\gamma\mathcal{L}\langle\pi^n_s,G_s\rangle|\leq C(G)$, therefore the limit in (28) is equal to zero. From Lemma $A1.5.1$ of [15], the expectation in (29) is equal to

$$\mathbb{E}_{\mu_n}\Big[\int_\tau^{\tau+\tilde{t}}n^\gamma\big[\mathcal{L}\langle\pi^n_s,G_s\rangle^2-2\langle\pi^n_s,G_s\rangle\mathcal{L}\langle\pi^n_s,G_s\rangle\big]\,ds\Big],$$

therefore to finish the proof it is enough to apply the next result.

Proposition 14 *Let $G\in\mathcal{S}$. Then*

$$\sup_{s\in[0,T]}\Big|n^\gamma\left(\mathcal{L}\langle\pi^n_s,G_s\rangle^2-2\langle\pi^n_s,G_s\rangle\mathcal{L}\langle\pi^n_s,G_s\rangle\right)\Big|\lesssim\max\{n^{\gamma-2},n^{-1}\}. \tag{30}$$

In particular,

$$\forall\delta_1>0,\quad\lim_{n\to\infty}\mathbb{P}_{\mu_n}\Big(\sup_{t\in[0,T]}|\mathcal{M}^n_t(G)|>\delta_1\Big)=0. \tag{31}$$

Proof We begin by showing (31) from (30). Since $\big(|\mathcal{M}^n_t(G)|\big)_{t\geq 0}$ is a non-negative submartingale, Doob's inequality leads to

$$\mathbb{P}_{\mu_n}\Big(\sup_{s\in[0,T]}|\mathcal{M}_s^n(G)|>\delta_1\Big)\leq\frac{1}{\delta_1}\Big(\mathbb{E}_{\mu_n}\Big[\big(\mathcal{M}_T^n(G)\big)^2\Big]\Big)^{1/2}$$
$$=\frac{1}{\delta_1}\Big(\int_0^T\sup_{s\in[0,T]}|n^\gamma\left(\mathcal{L}[\langle\pi_s^n,G_s\rangle]^2-2\langle\pi_s^n,G_s\rangle\mathcal{L}\langle\pi_s^n,G_s\rangle\right)|\,ds\Big)^{1/2},$$

which goes to zero as $n\to\infty$, for every $\delta_1>0$ fixed, due to (30). Now we go to the second part of the proof. After some algebraic manipulations, we rewrite the expression on the left-hand side of (30) as

$$\sup_{s\in[0,T]}\frac{n^\gamma}{4n^2}\sum_{x,y}[G_s\left(\tfrac{y}{n}\right)-G_s\left(\tfrac{x}{n}\right)]^2p(x-y)c_{x,y}^{(m)}(\eta_s^n)[\eta_s^n(x)-\eta_s^n(y)]^2.$$

Due to (5), we get $|c_{x,y}^{(m)}(\eta_s^n)|\leq 2m+1$. Combining this with Proposition A.10 of [8] and with the fact that $[\eta_s^n(x)-\eta_s^n(y)]^2\leq 1$, the proof ends. □

4.2 Characterization of Limit Points

From the results of Sect. 4.1, we know that $(\mathbb{Q}_n)_{n\geq1}$ has at least one limit point $\mathbb{Q}$. From Sect. 4.1 of [15], since every site has at most one particle, any limit point $\mathbb{Q}$ is concentrated on trajectories of measures that are absolutely continuous with respect to the Lebesgue measure, i.e.

$$\pi_t(du)=\rho(t,u)\,du, \tag{32}$$

for (almost) every t on $[0,T]$ with $0\leq\rho\leq1$ on $[0,T]\times\mathbb{R}$. In this subsection we will prove that $\mathbb{Q}$ is concentrated on trajectories such that ρ satisfies the first condition of weak solutions in Definition 5.

Proposition 15 *If $\mathbb{Q}$ is a limit point of $(\mathbb{Q}_n)_{n\geq1}$ then*

$$\mathbb{Q}\Big(\pi_\cdot\in\mathcal{D}\left([0,T],\mathcal{M}^+\right):\forall t\in[0,T],\forall G\in\mathcal{S},\quad F(t,\rho,G,g)=0\Big)=1,$$

where $F(t,\rho,G,g)$ is given in (8).

Last proposition identifies the profile $\rho(t,u)$ as a solution to the PDE.

Proof It is enough to verify that for any $\delta>0$ and any $G\in\mathcal{S}$,

$$\mathbb{Q}\Big(\pi_\cdot\in\mathcal{D}\left([0,T],\mathcal{M}^+\right):\ \sup_{t\in[0,T]}|F(t,\rho,G,g)|>\delta\Big)=0. \tag{33}$$

In order to simplify the notation, we will omit $\pi_\cdot$ from the sets where we are looking at. From the definition of F in Definition 5, last probability is bounded by

$$\mathbb{Q}\Big(\sup_{t\in[0,T]}\Big|\int_{\mathbb{R}}[\rho_t(u)G_t(u)-\rho_0(u)G_0(u)]\,du$$
$$-\int_0^t\int_{\mathbb{R}}\rho_s(u)\big[\big(\partial_s+[-(-\Delta)^{\gamma/2}]\big)G_s(u)\big]\,du\,ds\Big|>\frac{\delta}{2}\Big) \tag{34}$$
$$\mathbb{Q}\Big(\Big|\int_{\mathbb{R}}[\rho_0(u)-g(u)]\,G_0(u)\,du\Big|>\frac{\delta}{2}\Big). \tag{35}$$

The term in (35) is equal to zero, since $\mathbb{Q}$ is a limit point of $(\mathbb{Q}_n)_{n\geq 1}$ and $\mathbb{Q}_n$ is induced by $\mathbb{P}_{\mu_n}$, which is induced by μ_n, which is associated with g, see Definition 8. Then we are done if we can prove that the term in (34) is also equal to zero.

The proof for $m=1$ makes use of the same arguments applied for showing Proposition 4.2 in [8], therefore we omit it. The proof for $m\geq 2$ is analogous to the proof of Proposition 5.1 in [6], therefore we omit many steps. For $\varepsilon>0$ and $u\in\mathbb{R}$ fixed, we define $\overrightarrow{i_\varepsilon^u}$ by

$$\forall v\in\mathbb{R},\quad \overrightarrow{i_\varepsilon^u}(v):=\frac{1}{\varepsilon}\mathbb{1}_{(u,u+\varepsilon]}(v).$$

Putting this together with (32), we get $\langle\pi_s,\overrightarrow{i_\varepsilon^u}\rangle=\langle\rho_s,\overrightarrow{i_\varepsilon^u}\rangle=\frac{1}{\varepsilon}\int_u^{u+\varepsilon}\rho(s,v)dv$, and since $\rho\in[0,1]$ and Lebesgue's Differentiation Theorem, we conclude that

$$\lim_{\varepsilon\to 0^+}\Big|\prod_{j=0}^{m-1}\langle\pi_s,\overrightarrow{i_\varepsilon^{u+j\varepsilon}}\rangle-\rho^m(s,u)\Big|=0,$$

for almost every $u\in\mathbb{R}$. Therefore we are done if we can prove that

$$\overline{\lim_{\varepsilon\to 0^+}}\,\mathbb{Q}\Big(\sup_{t\in[0,T]}\Big|\langle\pi_t,G_t\rangle-\langle\pi_0,G_0\rangle-\int_0^t\langle\pi_s,\partial_s G_s\rangle\,ds$$
$$-\int_0^t\int_{\mathbb{R}}\prod_{j=0}^{m-1}\langle\pi_s,\overrightarrow{i_\varepsilon^{u+j\varepsilon}}\rangle[-(-\Delta)^{\gamma/2}G_s](u)\,du\,ds\Big|>\frac{\delta}{3}\Big). \tag{36}$$

We observe that the functions $\overrightarrow{i_\varepsilon^{u+j\varepsilon}}$ are not continuous; nevertheless, those functions can be approximated by continuous functions in such a way that the error vanishes as $\varepsilon\to 0^+$. Thus, by choosing convenient approximations and applying Portmanteau's Theorem (see Theorem 2.1 in [4]) in the same way as in the proof of Proposition 5.1 in [6], the display in (36) is bounded from above by

$$\varlimsup_{\varepsilon\to 0^+}\varlimsup_{n\to\infty}\mathbb{P}_{\mu_n}\Big(\sup_{t\in[0,T]}\Big|\int_0^t\frac{n^\gamma}{2n}\sum_x \mathcal{K}_n G_s(\tfrac{x}{n})\Big[\prod_{j=0}^{m-1}\eta_s^n(x-j)+\prod_{j=0}^{m-1}\eta_s^n(x+j)\Big]ds$$

$$+\mathcal{M}_t^n(G)+\int_0^t\mathcal{R}_n^G(s)ds-\int_0^t\int_{\mathbb{R}}[-(-\Delta)^{\gamma/2}G_s](u)\prod_{j=0}^{m-1}\langle\pi_s^n,\overrightarrow{i_\varepsilon^{u+j\varepsilon}}\rangle\,du\,ds\Big|>\frac{\delta}{4}\Big).$$

Above we applied (19) and (22), besides summing and subtracting $\int_0^t n^\gamma\mathcal{L}\langle\pi_s^n,G_s\rangle$ ds. Since the error from changing the integral in the space variable by its Riemann sum is or order n^{-1}, it is enough to prove that

$$\lim_{n\to\infty}\mathbb{P}_{\mu_n}\Big(\sup_{t\in[0,T]}\Big|\mathcal{M}_t^n(G)+\int_0^t\mathcal{R}_n^G(s)\,ds\Big|>\frac{\delta}{8}\Big)=0 \tag{37}$$

and

$$\varlimsup_{\varepsilon\to 0^+}\varlimsup_{n\to\infty}\mathbb{P}_{\mu_n}\Big(\sup_{t\in[0,T]}\Big|\int_0^t\Big\{\frac{1}{n}\sum_x n^\gamma[\mathcal{K}_n G_s(\tfrac{x+m-1}{n})+\mathcal{K}_n G_s(\tfrac{x}{n})]\prod_{j=0}^{m-1}\eta_s^n(x+j)$$

$$-\frac{1}{n}\sum_x[-(-\Delta)^{\gamma/2}G_s](\tfrac{x+m-1}{n})\prod_{j=0}^{m-1}\langle\pi_s^n,\overrightarrow{i_\varepsilon^{\frac{x}{n}+j\varepsilon}}\rangle$$

$$-\frac{1}{n}\sum_x[-(-\Delta)^{\gamma/2}G_s](\tfrac{x}{n})\prod_{j=0}^{m-1}\langle\pi_s^n,\overrightarrow{i_\varepsilon^{\frac{x}{n}+j\varepsilon}}\rangle\Big\}ds\Big|>\frac{\delta}{8}\Big)=0.$$

By combining (31) with (23) and Markov's inequality, we get (37). Here and in what follows we interpret εn as $\lfloor\varepsilon n\rfloor$. Putting together the estimate

$$\Big|\prod_{j=0}^{m-1}\langle\pi_s^n,\overrightarrow{i_\varepsilon^{\frac{x}{n}+j\varepsilon}}\rangle-\prod_{j=0}^{m-1}\overrightarrow{\eta}_s^{\varepsilon n}(x+j\varepsilon n)\Big|\lesssim(\varepsilon n)^{-1}$$

with (26) and the fact that $[-(-\Delta)^{\gamma/2}G_s]\in L^1(\mathbb{R})$, we are done if we can show that

$$\varlimsup_{\varepsilon\to 0^+}\varlimsup_{n\to\infty}\mathbb{P}_{\mu_n}\Big(\sup_{t\in[0,T]}\Big|\int_0^t\Big\{\frac{1}{n}\sum_x n^\gamma[\mathcal{K}_n G_s(\tfrac{x+m-1}{n})+\mathcal{K}_n G_s(\tfrac{x}{n})]\prod_{j=0}^{m-1}\eta_s^n(x+j)$$

$$-\frac{1}{n}\sum_x[[-(-\Delta)^{\gamma/2}G_s](\tfrac{x+m-1}{n})+[-(-\Delta)^{\gamma/2}G_s](\tfrac{x}{n})]\prod_{j=0}^{m-1}\overrightarrow{\eta}_s^{\varepsilon n}(x+j\varepsilon n)\Big\}ds\Big|>\frac{\delta}{8}\Big).$$

is equal to zero. Last display is bounded by the sum of the next three terms:

$$\varlimsup_{\varepsilon\to 0^+}\varlimsup_{n\to\infty}\mathbb{P}_{\mu_n}\Big(\sup_{t\in[0,T]}\Big|\int_0^t\frac{1}{n}\sum_x\Big|\Big(n^\gamma\mathcal{K}_n G_s-[-(-\Delta)^{\gamma/2}G_s]\Big)(\tfrac{x+m-1}{n})\Big|\,ds\Big|>\frac{\delta}{24}\Big),$$

$$\overline{\lim_{\varepsilon\to 0^+}}\lim_{n\to\infty}\mathbb{P}_{\mu_n}\Big(\sup_{t\in[0,T]}\Big|\int_0^t\frac{1}{n}\sum_x\big|n^\gamma\mathcal{K}_nG_s(\tfrac{x}{n})-[-(-\Delta)^{\gamma/2}G_s](\tfrac{x}{n})\big|\,ds\Big|>\frac{\delta}{24}\Big),$$

$$\overline{\lim_{\varepsilon\to 0^+}}\lim_{n\to\infty}\mathbb{P}_{\mu_n}\Big(\sup_{t\in[0,T]}\Big|\int_0^t\frac{1}{n}\sum_x\big([-(-\Delta)^{\gamma/2}G_s](\tfrac{x+m-1}{n})+[-(-\Delta)^{\gamma/2}G_s](\tfrac{x}{n})\big)$$
$$\times\Big[\prod_{j=0}^{m-1}\eta_s^n(x+j)-\prod_{j=0}^{m-1}\overrightarrow{\eta}_s^{\varepsilon n}(x+j\varepsilon n)\Big]ds\Big|>\frac{\delta}{24}\Big).$$

In the first and second terms of last display, we used the fact that $|\eta_s^n(\cdot)|\le 1$, for every $s\in[0,T]$. Combining Markov's inequality with Proposition 10, we conclude that they are equal to zero. Finally, since $[-(-\Delta)^{\gamma/2}G(s,\cdot)]\in L^1(\mathbb{R})\cap L^\infty(\mathbb{R})$ (due to Proposition 13), from Lemma 12, the last term in last display is equal to zero. This ends the proof. □

4.3 Energy Estimates

In this subsection we prove that $\mathbb{Q}$ is concentrated on trajectories of measures whose density ρ satisfies the second condition of weak solutions in Definition 5. We do so by deriving some *energy estimates*, in the sense that $\mathbb{Q}$ is concentrated in trajectories of measures such that a transformation of the density ρ given in (32) belongs to some Hilbert space. Here and in what follows, b is the constant given in Assumption (A). We do not prove the next result but we refer the reader to [7] for a proof. In what follows, we denote $Y:=L^2([0,T]\times\mathbb{R})$ and $\|\cdot\|_Y$ denotes the usual norm in Y.

Proposition 16 *Under Assumption (A), we have* $\mathbb{E}_{\mathbb{Q}}\big[\|\rho-b\|_Y\big]<\infty$.

Observe that since $b\in(0,1)$, we get

$$|\lambda^m-b^m|=|\lambda-b|\Big|\sum_{i=0}^{m-1}\lambda^i b^{m-1-i}\Big|\le|\lambda-b|\sum_{i=0}^{\infty}b^i=\frac{|\lambda-b|}{1-b},$$

for every $\lambda\in[0,1]$ and every $m\ge 1$. Therefore, the next result is a direct consequence of the previous result.

Corollary 17 *Under Assumption (A), there exists $K_2>0$ such that for all $m\ge 1$* $\mathbb{E}_{\mathbb{Q}}\big[\|\rho^m-b^m\|_Y\big]\le K_2$.

Now we characterize a limit point $\mathbb{Q}$ in a more detailed way. More precisely, we prove that a transformation of the density ρ given in (32) belongs to some fractional Sobolev space.

For the remainder of this subsection, we denote $N := C_c^{0,0,0}([0,T]\times\mathbb{R}^2)$. Moreover, for every $\epsilon\in(0,1]$, we define $O(\epsilon)$ and $Y(\epsilon)$ by

$$O(\epsilon) := \{(v,v)\in\mathbb{R}^2 : |u-v|\geq\epsilon\};\quad Y(\epsilon) := L^2([0,T]\times\mathbb{R}^2, dt\otimes d\mu_\epsilon), \tag{38}$$

where μ_ϵ is the measure $(u,v)\in\mathbb{R}^2\to\mathbb{1}_{\{|u-v|\geq\epsilon\}}|u-v|^{-1-\gamma}$. Denote the norm associated to the Hilbert space $Y(\epsilon)$ by $\|\cdot\|_{Y(\varepsilon)}$. Now we state a preliminary lemma that will be useful in what follows.

Lemma 18 *For every $\epsilon>0$, the metric space $(N,\|\cdot\|_{Y(\epsilon)})$ is dense in the metric space $\big(Y(\epsilon),\|\cdot\|_{Y(\epsilon)}\big)$ and the metric space $\big(Y(\epsilon),\|\cdot\|_{Y(\epsilon)}\big)$ is separable.*

We do not present the proof of last lemma since it is classical. Now, for $\epsilon\in(0,1]$, we define the (random) linear functional $\ell^{(m)}_{\rho,\epsilon}: N\to\mathbb{R}$ by

$$\forall H\in N,\quad \ell^{(m)}_{\rho,\epsilon}(H) := \int_0^T\iint_{O(\epsilon)}\frac{[\rho^m(t,v)-\rho^m(t,u)]H_t(u,v)}{|u-v|^{1+\gamma}}\,du\,dv\,dt.$$

Following the same strategy applied to show Theorem 3.2 i) in [3] it is possible to derive the next result.

Lemma 19 *Let $m\geq 1$. For every $\epsilon\in(0,1]$, let $O(\epsilon)$ and $Y(\epsilon)$ be given by* (38). *Moreover, assume that there exist positive constants K_0, K_1 such that*

$$\forall i\geq 1, \forall H_1,\ldots,H_i\in N,\quad \mathbb{E}_{\mathbb{Q}}\Big[\max_{1\leq\tilde{i}\leq i}\big\{\ell^{(m)}_{\rho,\epsilon}(H_{\tilde{i}})-K_0\|H_{\tilde{i}}\|^2_{Y(\epsilon)}\big\}\Big]\leq K_1. \tag{39}$$

Above, K_0 and K_1 are independent of i, $H_1,\ldots,H_i$ and ϵ. Furthermore, assume that $\mathbb{E}_{\mathbb{Q}}\big[\|\rho^m-b^m\|_Y\big]<\infty$. Then

$$\mathbb{E}_{\mathbb{Q}}\Big[\int_0^T\iint_{\mathbb{R}^2}\frac{[\rho^m(t,v)-\rho^m(t,u)]^2}{|u-v|^{1+\gamma}}\,du\,dv\,dt\Big]<\infty.$$

4.3.1 Useful Bounds

A particularly useful tool is the *Dirichlet form* defined by $\langle\sqrt{f},-n^\gamma\mathcal{L}\sqrt{f}\rangle_{\nu_b}$, where $f:\Omega\to\mathbb{R}$ is a density with respect to ν_b and for all functions $g,h:\Omega\to\mathbb{R}$, $\langle g,h\rangle_{\nu_b}$ denotes the scalar product in $L^2(\Omega,\nu_b)$. Now we define $\mathfrak{D}(\sqrt{f},\nu_b)$ as

$$\mathfrak{D}(\sqrt{f},\nu_b) := \frac{1}{4}\sum_{x,y}p(y-x)I_{x,y}(\sqrt{f},\nu_b). \tag{40}$$

where for every $x,y\in\mathbb{Z}$

$$I_{x,y}(\sqrt{f},\nu_b) := \int_\Omega c^{(m)}_{x,y}(\eta)\big[\sqrt{f(\eta^{x,y})}-\sqrt{f(\eta)}\big]^2\,d\nu_b, \tag{41}$$

and $c^{(m)}_{x,y}(\eta)$ is given in (5). From (6) it is easy to check that

$$\langle \mathcal{L}\sqrt{f},\sqrt{f}\rangle_{\nu_b}=-\frac{1}{2}\mathfrak{D}(\sqrt{f},\nu_b), \tag{42}$$

for any density f. Finally, we are ready to prove that ρ satisfies the second condition of weak solutions in Definition 5. We present the proof only for the case $m\geq 2$, since for $m=1$ it is enough to apply Proposition 5.2 of [8].

Proposition 20 *Assume* (9) *and* $m\geq 2$. *Then*

$$\mathbb{Q}\Big(\rho^m-b^m\in L^2\big(0,T;\mathcal{H}^{\gamma/2}(\mathbb{R})\big)\Big)=1.$$

Proof From Proposition 16 it is enough to prove that

$$\mathbb{Q}\Big(\int_0^T\iint_{\mathbb{R}^2}\frac{[\rho^m_t(v)-\rho^m_t(u)]^2}{|u-v|^{1+\gamma}}du\,dv\,dt<\infty\Big)=1.$$

We do so by applying Lemma 19. From Corollary 17, we are done if we can prove (39). With this in mind, fix $\epsilon\in(0,1]$, $j\geq 1$ and $G_1,\ldots,G_j\in N$. Now for every ε, $K_0>0$, define $\Psi_{K_0,\varepsilon}(\pi):=\max_{1\leq i\leq j}\big\{-K_0\|G_i\|_{Y(\epsilon)}+I_\varepsilon(\pi,G_i)\big\}\in\mathbb{R}$, for any $\pi\in\mathcal{D}([0,T],\mathcal{M}^+)$, where

$$I_\varepsilon(\pi,G_i):=\int_0^T\iint_{O(\epsilon)}\frac{\Big[\prod_{j=0}^{m-1}\langle\pi_t,\overrightarrow{i_\varepsilon^{v+j\varepsilon}}\rangle-\prod_{j=0}^{m-1}\langle\pi_t,\overrightarrow{i_\varepsilon^{u+j\varepsilon}}\rangle\Big]G_i(t,u,v)}{|u-v|^{1+\gamma}}du\,dv\,dt.$$

Following [6], since $0\leq\rho\leq 1$ and $G_1,\ldots,G_j$ are fixed, we have from (32) and Lebesgue's Differentiation Theorem that

$$\varlimsup_{\varepsilon\to 0^+}\mathbb{E}_{\mathbb{Q}}\big[\max_{1\leq i\leq j}\{\ell^{(m,\epsilon)}_\rho(G_i)-I_\varepsilon(\pi,G_i)\}\big]=0.$$

Therefore, in order to obtain (39), it is enough to prove that there exist K_0, K_1 such that $\varlimsup_{\varepsilon\to 0^+}\mathbb{E}_{\mathbb{Q}}[\Psi_{K_0,\varepsilon}(\pi)]\leq K_1$. Next for every $K_0,\varepsilon>0$, every $n\geq 1$ and every $G\in N$, define the random application $\tilde{Z}^{K_0,\varepsilon}_{n,G}:\mathcal{D}([0,T],\Omega)\to\mathbb{R}$ by

$$\tilde{Z}^{K_0,\varepsilon}_{n,G}:=\int_0^T\Big\{n^{\gamma-1}\sum_{(x,y)\in I^n(\epsilon)}\Big[\prod_{j=0}^{m-1}\overrightarrow{\eta}^{\varepsilon n}_t(y+j\varepsilon n)-\prod_{j=0}^{m-1}\overrightarrow{\eta}^{\varepsilon n}_t(x+j\varepsilon n)\Big]\times$$
$$\times G(t,\tfrac{x}{n},\tfrac{y}{n})|y-x|^{-\gamma-1}\Big\}dt-K_0\|G\|_{Y(\epsilon)},$$

where $I^n(\epsilon):=\{(x,y)\in\cap\mathbb{Z}^2:|x-y|\geq\epsilon n\}$. Since $\mathbb{Q}$ is a limit point, there exists a subsequence $(\mathbb{Q}_{n'})_{n'\geq 1}$ of $(\mathbb{Q}_n)_{n\geq 1}$ converging weakly to $\mathbb{Q}$. Combining this with the fact that $\mathbb{Q}_n$ is induced by $\mathbb{P}_{\mu_n}$ for every $n\geq 1$ and that $\Psi_{K_0,\varepsilon}$ is bounded and lower semi-continuous with respect to the Skorohod topology of $\mathcal{D}([0,T],\mathcal{M}^+)$ for every $K_0, \varepsilon > 0$, we get

$$\overline{\lim_{\varepsilon\to 0^+}}\ \mathbb{E}_{\mathbb{Q}}[\Psi_{K_0,\varepsilon}(\pi)] = \overline{\lim_{\varepsilon\to 0^+}}\ \lim_{n'\to\infty} \mathbb{E}_{\mathbb{Q}_{n'}}[\Psi_{K_0,\varepsilon}(\pi)]$$
$$= \overline{\lim_{\varepsilon\to 0^+}}\ \lim_{n'\to\infty} \mathbb{E}_{\mu_{n'}}\Big[\max_{1\le i\le j}\{\tilde{Z}^{K_0}_{n',G_i}\}\Big] \le \overline{\lim_{\varepsilon\to 0^+}}\ \overline{\lim_{n\to\infty}}\ \mathbb{E}_{\mu_n}\Big[\max_{1\le i\le j}\{\tilde{Z}^{K_0}_{n,G_i}\}\Big].$$

This means that we are done if we can prove that

$$\exists K_0, K_1 > 0: \quad \overline{\lim_{\varepsilon\to 0^+}}\ \overline{\lim_{n\to\infty}}\ \mathbb{E}_{\mu_n}\Big[\max_{1\le i\le j}\{\tilde{Z}^{K_0}_{n,G_i}\}\Big] \le K_1. \tag{43}$$

We do so by applying the strategy in the proof of Proposition 5.3 in [6]. For every $G \in N$, we define $(\Phi^G_n)_{n\ge 1} : [0,T]\times\mathbb{R}\to\mathbb{R}$ by

$$\forall t \in [0,T], \forall x \in \mathbb{Z}, \quad \Phi^G_n(t, \tfrac{x}{n}) := n^{\gamma} \sum_{y:|x-y|\ge \epsilon n} G(t, \tfrac{x}{n}, \tfrac{y}{n})|y-x|^{-\gamma-1}.$$

Following the arguments in [6], we have that $(\Phi_n)^G_{n\ge 1}$ satisfies (25). From Lemma 12, for every $G \in N$, it holds

$$\overline{\lim_{\varepsilon\to 0^+}}\ \overline{\lim_{n\to\infty}}\ \mathbb{E}_{\mu_n}\Big[\int_0^T n^{\gamma-1} \sum_{x,y:|x-y|\ge \epsilon n} A_{x,y}(\varepsilon n, \eta^n_t) G(t, \tfrac{x}{n}, \tfrac{y}{n})|y-x|^{-\gamma-1}\, dt\Big] = 0,$$

where for every $\eta \in \Omega$, any $x, y \in \mathbb{Z}$ and every $\ell \ge 1$, $A_{x,y}(\ell, \eta)$ is given by

$$A_{x,y}(\ell,\eta) := \prod_{j=0}^{m-1} \overrightarrow{\eta}^{\ell}(x+j\ell) - \prod_{j=0}^{m-1} \eta(x+j) + \prod_{j=0}^{m-1} \overrightarrow{\eta}^{\ell}(y+j\ell) - \prod_{j=0}^{m-1} \eta(y+j).$$

for any $G \in N$. Then in order to get (43), it is enough to find $K_0, K_1 > 0$ such that

$$\overline{\lim_{n\to\infty}}\ \mathbb{E}_{\mu_n}\Big[\max_{1\le i\le j}\{Z^{K_0}_{n,G_i}\}\Big] \le K_1, \tag{44}$$

where for any $K_0, n \ge 1$ and $G \in N$, $Z^{K_0}_{n,G} : \mathcal{D}([0,T],\Omega) \to \mathbb{R}$ is given by

$$Z^{K_0}_{n,G} := \int_0^T \Big\{ n^{\gamma-1} \sum_{(x,y)\in I^n(\epsilon)} \tilde{A}_{x,y}(\eta^n_t) G(t, \tfrac{x}{n}, \tfrac{y}{n})|y-x|^{-\gamma-1} - \frac{K_0\|G\|^2_{Y(\epsilon)}}{T}\Big\}\, dt.$$

Above, for any $\eta \in \Omega$ and every $x, y \in \mathbb{Z}$, $\tilde{A}_{x,y}(\eta)$ is given by

$$\tilde{A}_{x,y}(\eta) := \prod_{j=0}^{m-1} \eta(y+j) - \prod_{j=0}^{m-1} \eta(x+j).$$

Combining entropy and Jensen's inequalities with the inequality

$$\overline{\lim_{n\to\infty}} \left(\max\{a_n, b_n\}\right)_{n\geq 1} \leq \max\left\{ \overline{\lim_{n\to\infty}} (a_n)_{n\geq 1}, \overline{\lim_{n\to\infty}} (b_n)_{n\geq 1}\right\},$$

we get the following estimate.

$$\overline{\lim_{n\to\infty}} \mathbb{E}_{\mu_n}\big[\max_{1\leq i\leq j} \{Z^{K_0}_{n,G_i}\}\big] \leq C_b + \max_{1\leq i\leq j} \overline{\lim_{n\to\infty}} \frac{1}{n} \log\Big(\mathbb{E}_{\nu_b}\big[\exp(nZ^{K_0}_{n,G_i})\big]\Big).$$

Therefore, in order to show that (44) holds for $K_1 = C_b$ and end the proof, it is enough to prove the following claim:

$$\exists K_0 > 0: \quad \sup_{G\in N} \overline{\lim_{n\to\infty}} \frac{1}{n} \log\Big(\mathbb{E}_{\nu_b}\big[\exp(nZ^{K_0}_{n,G})\big]\Big) \leq 0. \tag{45}$$

Keeping this in mind, we apply Feynman-Kac's formula (see Lemma A.1 in [1]) and bound the last $\overline{\lim}$ from above by

$$\begin{aligned} \overline{\lim_{n\to\infty}} \int_0^T \sup_f \Big\{ &\sum_{x,y:|x-y|\geq \epsilon n} \frac{n^{\gamma-1}}{|y-x|^{\gamma+1}} \int_\Omega \tilde{A}_{x,y}(\eta) G_t(\tfrac{x}{n}, \tfrac{y}{n}) f(\eta)\, d\nu_b \\ &+ n^{\gamma-1} \langle \mathcal{L}\sqrt{f}, \sqrt{f}\rangle_{\nu_b} - \frac{K_0 \|G\|^2_{Y(\epsilon)}}{T} \Big\}\, dt, \end{aligned} \tag{46}$$

where last supremum is carried over all the densities f with respect to ν_b. Recall (5); in order to apply (42), the sum inside the supremum in (46) will be given in terms of

$$\sum_{x,y:|x-y|\geq \epsilon n} \frac{n^{\gamma-1}}{|y-x|^{\gamma+1}} G_t(\tfrac{x}{n}, \tfrac{y}{n}) \int_\Omega [\eta(y) - \eta(x)] c^{(m)}_{x,y}(\eta) f(\eta)\, d\nu_b.$$

For $n \geq 2\epsilon^{-1}$, $\varepsilon n \geq 2$ and $c^{(m)}_{x,y}(\eta) = c^{(m,dif)}_{x,y}(\eta) + c^{(m,dif)}_{y,x}(\eta)$ when $|x-y| \geq \epsilon n$. Next define $\hat{A}_m(\eta, x, y) := C_m(\eta, y, x) - C_m(\eta, x, y)$ for every $\eta \in \Omega$ and every $x, y \in \mathbb{Z}$, where $C_m(\cdot, \cdot, \cdot)$ is given in (18). Then from (16), the sum inside the supremum in (46) is equal to

$$\sum_{x,y:|x-y|\geq \epsilon n} \frac{n^{\gamma-1}}{2|y-x|^{\gamma+1}} G_t(\tfrac{x}{n},\tfrac{y}{n}) \int_\Omega [\eta(y)-\eta(x)] c^{(m)}_{x,y}(\eta)\, d\nu_b$$
$$+ \sum_{x,y:|x-y|\geq \epsilon n} \frac{n^{\gamma-1}}{2|y-x|^{\gamma+1}} [G_t(\tfrac{x}{n},\tfrac{y}{n}) - G_t(\tfrac{x+m-1}{n},\tfrac{y+m-1}{n})] \int_\Omega \tilde{A}_{x,y}(\eta) f(\eta)\, d\nu_b$$
$$+ \sum_{x,y:|x-y|\geq \epsilon n} \frac{n^{\gamma-1}}{2|y-x|^{\gamma+1}} [G_t(\tfrac{x-1}{n},\tfrac{y-1}{n}) - G_t(\tfrac{x}{n},\tfrac{y}{n})] \int_\Omega \hat{A}_m(\eta,x,y) f(\eta)\, d\nu_b.$$

In particular, since f is a density with respect to ν_b and there exists $b_G > 0$ such that $G(t,u,v) = 0$ when $|u| \geq b_G$ or $|v| \geq b_G$, the sum of the terms in the second and third lines of last display is bounded from above by

$$mC(G,m,n) n^{\gamma-1} \sum_{x,y:|x-y|\geq \epsilon n, |x|+|y|\leq 4b_G n} |y-x|^{-\gamma-1} \lesssim m b_G \epsilon^{-\gamma} C(G,m,n),$$

where $C(G,m,n)$ is given by

$$C(G,m,n) := \sup_{t\in[0,T], \{(u_2-u_1)^2+(v_2-v_1)^2 < 2m^2 n^{-2}\}} |G_t(u_2,v_2) - G_t(u_1,v_1)|.$$

Since G is uniformly continuous we get $\lim_{n\to\infty} C(G,m,n) = 0$. Therefore we conclude that the display in (46) is equal to

$$\overline{\lim_{n\to\infty}} \int_0^T \sup_f \Big\{ \sum_{(x,y)\in I^n(\epsilon)} \frac{n^{\gamma-1}}{2|y-x|^{\gamma+1}} G_t(\tfrac{x}{n},\tfrac{y}{n}) \int_\Omega [\eta(y)-\eta(x)] c^{(m)}_{x,y}(\eta)\, d\nu_b$$
$$+ n^{\gamma-1} \langle \mathcal{L}\sqrt{f}, \sqrt{f}\rangle_{\nu_b} - \frac{K_0 \|G\|^2_{Y(\epsilon)}}{T} \Big\}\, dt. \tag{47}$$

Next we bound the sum inside the supremum in (47) from above by

$$\sum_{(x,y)\in I^n(\epsilon)} \frac{n^{\gamma-1}}{2|y-x|^{\gamma+1}} |G_t(\tfrac{x}{n},\tfrac{y}{n})| \Big| \int_\Omega [\eta(y)-\eta(x)] c^{(m)}_{x,y}(\eta) f(\eta)\, d\nu_b \Big|.$$

From (6), the fact that $c^{(m)}_{x,y}(\eta) = c^{(m)}_{y,x}(\eta)$, Young's inequality and (41), we get

$$\Big| \int_\Omega [\eta(y)-\eta(x)] c^{(m)}_{x,y}(\eta) f(\eta)\, d\nu_b \Big|$$
$$\leq \frac{1}{2} \int_\Omega c^{(m)}_{x,y}(\eta) |\sqrt{f(\eta)} - \sqrt{f(\eta^{x,y})}| [\sqrt{f(\eta)} + \sqrt{f(\eta^{x,y})}]\, d\nu_b$$
$$\leq \frac{m}{E_{x,y}} \int_\Omega [f(\eta) + f(\eta^{x,y})]\, d\nu_b + \frac{E_{x,y}}{4} I_{x,y}(\sqrt{f}, \nu_b),$$

for every $x, y \in \mathbb{Z}$ and every $E_{x,y} > 0$. Above, we used the fact (which is due to (5) and Remark 2) that $c_{x,y}^{(m)} \leq 2m$. From (6) and the fact that f is a density with respect to ν_b, last display can be rewritten as

$$\frac{2m}{E_{x,y}} + \frac{E_{x,y}}{4} I_{x,y}(\sqrt{f}, \nu_b),$$

therefore the sum inside the supremum in (47) is bounded from above by

$$\sum_{(x,y)\in I^n(\epsilon)} \frac{n^{\gamma-1}}{2|y-x|^{\gamma+1}} |G_t(\tfrac{x}{n}, \tfrac{y}{n})| \Big(\frac{E_{x,y}}{4} I_{x,y}(\sqrt{f}, \nu_b) + \frac{2m}{E_{x,y}} \Big). \tag{48}$$

From (42), we rewrite the second term inside the supremum in (47) as

$$-\frac{n^{\gamma-1}}{2}\mathscr{D}(\sqrt{f}, \nu_b) \leq -\frac{n^{\gamma-1}}{8} \sum_{(x,y)\in I^n(\epsilon)} c_\gamma |y-x|^{-1-\gamma} I_{x,y}(\sqrt{f}, \nu_b).$$

Combining last display with (48), we conclude that the expression inside the supremum in (47) is bounded from above by

$$\frac{n^{\gamma-1}}{8} \sum_{(x,y)\in I^n(\epsilon)} |y-x|^{-1-\gamma} I_{x,y}(\sqrt{f}, \nu_b) \Big[E_{x,y} |G_t(\tfrac{x}{n}, \tfrac{y}{n})| - c_\gamma \Big]$$
$$- \frac{K_0 \|G\|^2_{Y(\epsilon)}}{T} + n^{\gamma-1} \sum_{(x,y)\in I^n(\epsilon)} |G_t(\tfrac{x}{n}, \tfrac{y}{n})| |y-x|^{-1-\gamma} \frac{m}{E_{x,y}}.$$

Choosing $E_{x,y} = c_\gamma (|G_t(\frac{x}{n}, \frac{y}{n})|)^{-1}$, last display can be rewritten as

$$- \frac{K_0 \|G\|^2_{Y(\epsilon)}}{T} + \frac{m}{c_\gamma} n^{\gamma-1} \sum_{(x,y)\in I^n(\epsilon)} [G_t(\tfrac{x}{n}, \tfrac{y}{n})]^2 |y-x|^{-1-\gamma},$$

therefore the display in (47) is bounded from above by

$$\begin{aligned}
&\varlimsup_{n\to\infty} \int_0^T \Big\{ - \frac{K_0 \|G\|^2_{Y(\epsilon)}}{T} + \frac{m}{c_\gamma} \frac{1}{n^2} \sum_{(x,y)\in I^n(\epsilon)} [G_t(\tfrac{x}{n}, \tfrac{y}{n})]^2 \Big| \frac{y-x}{n} \Big|^{-1-\gamma} \Big\} dt \\
&= - K_0 \|G\|^2_{Y(\epsilon)} + \frac{m}{c_\gamma} \int_0^T \iint_{O(\epsilon)} \frac{[G_t(\frac{x}{n}, \frac{y}{n})]^2}{|u-v|^{1+\gamma}} du\, dv\, dt \\
&= \Big(\frac{m}{c_\gamma} - K_0 \Big) \|G\|^2_{Y(\epsilon)}.
\end{aligned}$$

Finally, by choosing $K_0 = m(c_\gamma)^{-1}$, we get (45) and the proof ends. □

4.4 Replacement Lemmas

In this subsection, we present the proof of Lemma 12 (we refer the interested reader to Sect. 6 of [6] for more details). Let $\varepsilon > 0$ and $n \geq 1$ be such that $\varepsilon n^{\gamma/2} \geq 1$. In the first step, similarly to Lemma 5.3 in [5] and Lemma 6.2 in [6], only jumps of size 1 are used to replace products of η_s^n's by products of empirical averages in boxes of size $\ell = \varepsilon n^{\gamma/2}$. The latter products are then replaced by products of empirical averages in boxes of size $L = \varepsilon n$ in the second step, by applying long jumps. It is crucial that our dynamics always allows exchanges of particles between neighbor sites. For $m = 2$, this is assured by Remark 1 (this remark was applied often in Sect. 6 of [6]). For $m \geq 3$, this is assured by (5).

Next we illustrate the strategy for the first step. Given $\eta \in \Omega$ and $x \in \mathbb{Z}$, observe that for $m = 2$, it holds

$$\begin{aligned}\prod_{i=0}^{m-1} \eta(x+i) - \prod_{i=0}^{m-1} \overrightarrow{\eta}^{\ell}(x+i\ell) &= \eta(x)\eta(x+1) - \overrightarrow{\eta}^{\ell}(x)\overrightarrow{\eta}^{\ell}(x+\ell)\\ &= \eta(x)[\eta(x+1) - \overrightarrow{\eta}^{\ell}(x+\ell)] + \overrightarrow{\eta}^{\ell}(x+\ell)[\eta(x) - \overrightarrow{\eta}^{\ell}(x)].\end{aligned}$$

More generally, for every $m \geq 2$, it holds

$$\prod_{i=0}^{m-1} \eta(x+i) - \prod_{i=0}^{m-1} \overrightarrow{\eta}^{\ell}(x+i\ell) = \sum_{j=1}^{m} B^{\ell}_{x,j}(\eta)\big[\eta(x+m-j) - \overrightarrow{\eta}^{\ell}\big(x+(m-j)\ell\big)\big],$$

where for every $x \in \mathbb{Z}$, $\eta \in \Omega$ and $j \in \{1, \dots, m\}$, $B^{\ell}_{x,j}(\eta)$ is defined by

$$B^{\ell}_{x,j}(\eta) := \prod_{i=0}^{m-j-1} \eta(x+i) \prod_{i=m-j+1}^{k-1} \overrightarrow{\eta}^{\ell}(x+i\ell). \tag{49}$$

This motivates us to state next result.

Lemma 21 *Let $t \in [0, T]$. Assume $(\Phi_n)_{n\geq 1} : [0, T] \times \mathbb{R} \to \mathbb{R}$ satisfies (25) and denote $\ell = \varepsilon n^{\gamma/2}$. Then for every $j \in \{1, \dots, m\}$, it holds*

$$\overline{\lim_{\varepsilon\to 0^+}}\ \overline{\lim_{n\to\infty}}\ \mathbb{E}_{\mu_n}\Big[\Big|\int_0^t \sum_x \Phi_n(s, \tfrac{x}{n})\frac{B^{\ell}_{x,j}(\eta^n_s)}{n}\big[\eta^n_s(x+m-j) - \overrightarrow{\eta}^{\ell}_s\big(x+(m-j)\ell\big)\big]\,ds\Big|\Big]=0.$$

In particular, we have

$$\overline{\lim_{\varepsilon\to 0^+}}\ \overline{\lim_{n\to\infty}}\ \mathbb{E}_{\mu_n}\Big[\Big|\int_0^t \frac{1}{n}\sum_x \Phi_n(s, \tfrac{x}{n})\Big(\prod_{i=0}^{m-1} \eta^n_s(x+i) - \prod_{i=0}^{m-1} \overrightarrow{\eta}^{\ell}_s(x+i\ell)\Big)\,ds\Big|\Big] = 0.$$

Proof Combining entropy and Jensen's inequalities with Feynman-Kac's formula, we are done if we can prove that, for every $j \in \{1, \dots, m\}$,

$$\sup_f \Big\{ \Big| \frac{1}{n} \sum_x \Phi_n(s, \tfrac{x}{n}) \int_\Omega B^\ell_{x,j}(\eta)\big[\eta(x+m-j) - \overrightarrow{\eta}^\ell\big(x+(m-j)\ell\big)\big]\, d\nu_b \Big| \\ + n^{\gamma-1}\varepsilon \langle \mathcal{L}\sqrt{f}, \sqrt{f}\rangle_{\nu_b} \Big\}, \tag{50}$$

goes to zero, when we take first the lim sup for $n \to \infty$ and afterwards the lim sup for $\varepsilon \to 0^+$. In (50), the supremum is carried over all densities f with respect to ν_b. In order to treat the first term inside this supremum, we observe that

$$\eta(x+m-j) - \overrightarrow{\eta}^\ell\big(x+(m-j)\ell\big) = \frac{1}{\ell} \sum_{y=1}^{\ell} \sum_{z=x+m-j}^{x+(m-j)\ell+y-1} [\eta(z) - \eta(z+1)].$$

Therefore, the first term inside the supremum in (50) is bounded from above by

$$\Big| \sum_x \Phi_n(s, \tfrac{x}{n}) \sum_{y=1}^{\ell} \sum_{z=x+m-j}^{x+(m-j)\ell+y-1} \int_\Omega \frac{B^\ell_{x,j}(\eta)}{n\ell} [\eta(z) - \eta(z+1)] f(\eta)\, d\nu_b \Big|. \tag{51}$$

From (49), we have that $B_{x,j}(\eta) = B_{x,j}(\eta^{z,z+1})$ when $x+m-j \leq z \leq x+(m-j)\ell + y - 1$ and $1 \leq y \leq \ell$, for every $x \in \mathbb{Z}$, $j \in \{1, \dots, m\}$ and $\eta \in \Omega$. Then, by writing $f(\eta) = \frac{1}{2} f(\eta) + \frac{1}{2} f(\eta)$ and performing the change of variables $\eta \to \eta^{z,z+1}$, the display in (51) is equal to

$$\Big| \sum_x \Phi_n(s, \tfrac{x}{n}) \sum_{y=1}^{\ell} \sum_{z=x+m-j}^{x+(m-j)\ell+y-1} \int_\Omega \frac{B^\ell_{x,j}(\eta)}{2n\ell} [\eta(z) - \eta(z+1)][f(\eta) - f(\eta^{z,z+1})]\, d\nu_b \Big|.$$

By Young's inequality and (25), last display is bounded from above by

$$\frac{m\ell}{A} M_1 + \frac{AM_2}{4n\ell p(1)} \sum_x \sum_{y=1}^{\ell} \sum_{z=x+m-j}^{x+(m-j)\ell+y-1} p(1) I_{z,z+1}(\sqrt{f}, \nu_b). \tag{52}$$

for any $A > 0$. Above, we used the facts that $[B_{x,j}(\eta)]^2 [\eta(z) - \eta(z+1)]^2 \leq 1$ (due to (49) and $\eta(\cdot) \leq 1$) and f is a density with respect to ν_b. In order to apply (42), it is crucial to count the maximum number of times that a fixed bond $\{z_0, z_0+1\}$ appears in the triple sum in (52).

Observe that for $x \leq z_0 - m\ell$ or $x > z_0$ this bond does not appear. And for every $x \in \{z_0 - m\ell + 1, \dots, z_0\}$, $\{z_0, z_0+1\}$ is counted at most ℓ times (since y ranges over $\{1, \ell\}$). Hence every bond $\{z_0, z_0+1\}$ is counted at most $4m\ell^2$ times in the triple sum in (52). Therefore, from (40) and (42), the expression inside the supremum in (50) is bounded from above by

$$\frac{m\ell}{A}M_1+\Big[A\frac{M_2 m\ell}{4np(1)}-\frac{n^\gamma}{n}\varepsilon\Big]\mathfrak{D}(\sqrt{f},\nu_b),$$

for any $A>0$. Since $\ell=\varepsilon n^{\gamma/2}$, by choosing $A=4p(1)n^\gamma\varepsilon(M_2 m\ell)^{-1}$, last display is bounded from above by

$$\frac{M_1M_2m^2}{4p(1)}\frac{\ell^2}{n^\gamma\varepsilon}\lesssim\frac{\ell^2}{n^\gamma\varepsilon}=\frac{(\varepsilon n^{\gamma/2})^2}{n^\gamma\varepsilon}=\varepsilon,$$

which vanishes as $\varepsilon\to0^+$, and the proof ends. □

Next we state an auxiliary lemma which will be useful in the proof of Lemma 23. We omit its proof, since the strategy is analogous to the one used to show Lemma 6.3 in [6].

Lemma 22 (Moving Particle Lemma) *Fix $r\neq0\in\mathbb{Z}$ and f a density with respect to ν_b on Ω. For every $x\in\mathbb{Z}$, let $\Omega_x:=\{\eta\in\Omega:\eta(x+j)=1,\ j=1,2,\ldots,k\}$. Then*

$$\sum_x\int_{\Omega_x}\big[\sqrt{f(\eta^{x,x+r})}-\sqrt{f(\eta)}\big]^2d\nu_b\lesssim|r|^\gamma\mathfrak{D}(\sqrt{f},\nu_b).$$

Now, we illustrate the strategy for the second step. For every $m\geq2$, it holds

$$\prod_{i=0}^{m-1}\overrightarrow{\eta}^{\ell}(x+i\ell)-\prod_{i=0}^{m-1}\overrightarrow{\eta}^{\varepsilon n}(x+i\varepsilon n)$$
$$=\sum_{j=1}^{m}\tilde{B}^{\ell}_{x,j}(\eta)\big[\overrightarrow{\eta}^{\ell}\big(x+(m-j)\ell\big)-\overrightarrow{\eta}^{\varepsilon n}\big(x+(m-j)\varepsilon n\big)\big],$$

where $\ell\leq\varepsilon n$ and for every $x\in\mathbb{Z}$, $\eta\in\Omega$ and $j\in\{1,\ldots,m\}$, $\tilde{B}^{\ell}_{x,j}(\eta)$ is defined by

$$\tilde{B}^{\ell}_{x,j}(\eta):=\prod_{i=0}^{m-j-1}\overrightarrow{\eta}^{\ell}(x+i\ell)\prod_{i=m-j+1}^{k-1}\overrightarrow{\eta}^{\varepsilon n}(x+i\varepsilon n).\tag{53}$$

This motivates us to state next result. The proof is analogous to the proof of Lemma 6.5 in [6], therefore we omit some steps.

Lemma 23 *Assume $(\Phi_n)_{n\geq1}:[0,T]\times\mathbb{R}\to\mathbb{R}$ satisfies (25) and denote $\ell=\varepsilon n^{\gamma/2}$. Then for every $j\in\{1,\ldots,m\}$, it holds*

$$\varlimsup_{\varepsilon\to0^+}\varlimsup_{n\to\infty}\mathbb{E}_{\mu_n}\Big[\Big|\int_0^t\frac{1}{n}\sum_x\Phi_n(s,\tfrac{x}{n})\tilde{B}^{\ell}_{x,j}(\eta^n_s)$$
$$\times\big[\overrightarrow{\eta}^{\ell}_s\big(x+(m-j)\ell\big)-\overrightarrow{\eta}^{\varepsilon n}_s\big(x+(m-j)\varepsilon n\big)\big]ds\Big|\Big]=0.$$

In particular, we have

$$\overline{\lim_{\varepsilon\to 0^+}}\ \overline{\lim_{n\to\infty}}\ \mathbb{E}_{\mu_n}\Big[\Big|\int_0^t \frac{1}{n}\sum_x \Phi_n(s,\tfrac{x}{n})\Big[\prod_{i=0}^{m-1}\overrightarrow{\eta}_s^{\ell}(x+i\ell)-\prod_{i=0}^{m-1}\overrightarrow{\eta}_s^{\varepsilon n}(x+i\varepsilon n)\Big]ds\Big|\Big]=0.$$

Proof By applying the initial steps of the proof of Lemma 21, we are done if we can prove that, for every $j\in\{1,\dots,m\}$,

$$\sup_f\Big\{\Big|\int_\Omega \frac{1}{n}\sum_x \Phi_n(s,\tfrac{x}{n})\tilde{B}^{\ell}_{x,j}(\eta)\big[\overrightarrow{\eta}^{\ell}(x+(m-j)\ell)-\overrightarrow{\eta}^{\varepsilon n}(x+(m-j)\varepsilon n)\big]d\nu_b\Big|$$
$$+n^{\gamma-1}\varepsilon^{\gamma/2}\langle\mathcal{L}\sqrt{f},\sqrt{f}\rangle_{\nu_b}\Big\},\tag{54}$$

goes to zero, when we take first the lim sup for $n\to\infty$ and afterwards the lim sup for $\varepsilon\to 0^+$. In (54), the supremum is carried over all densities f with respect to ν_b. In order to treat the first term inside this supremum, we write $\varepsilon n = k\ell$ and observe that

$$\begin{aligned}&\overrightarrow{\eta}^{\ell}(x+(m-j)\ell)-\overrightarrow{\eta}^{\varepsilon n}(x+(m-j)\varepsilon n)\\ =&\frac{1}{k\ell}\sum_{i=0}^{k-1}\sum_{y=1}^{\ell}\big[\eta\big(x+(m-j)\ell+y\big)-\eta\big(x+(m-j)\ell+y+\ell[i+(m-j)(k-1)]\big)\big]\\ =&\frac{1}{k\ell}\sum_{i=0}^{k-1}\sum_{y=1}^{\ell}[\eta(z_{x,y})-\eta(z_{x,y}+\ell_i)],\end{aligned}$$

with $z_{x,y}:=x+(m-j)\ell+y$ and $\ell_i:=\ell[i+(m-j)(k-1)]$. Thus, the first term inside the supremum in (54) is equal to

$$\Big|\sum_x \Phi_n(s,\tfrac{x}{n})\sum_{i=0}^{k-1}\sum_{y=1}^{\ell}\int_\Omega \frac{\tilde{B}^{\ell}_{x,j}(\eta)}{k\ell n}[\eta(z_{x,y})-\eta(z_{x,y}+\ell_i)]f(\eta)\,d\nu_b\Big|.\tag{55}$$

From (53), we have that $\tilde{B}^{\ell}_{x,j}(\eta)=\tilde{B}^{\ell}_{x,j}(\eta^{z_{x,y},z_{x,y}+\ell_i})$ when $0\le i\le k-1$ and $1\le y\le\ell$, for every $x\in\mathbb{Z}$, $j\in\{1,\dots,m\}$ and $\eta\in\Omega$. Then, by writing $f(\eta)=\frac{1}{2}f(\eta)+\frac{1}{2}f(\eta)$ and performing the change of variables $\eta\to\eta^{z_{x,y},z_{x,y}+\ell_i}$, the display in (55) is equal to

$$\Big|\sum_x \Phi_n(s,\tfrac{x}{n})\sum_{i=0}^{k-1}\sum_{y=1}^{\ell}\int_\Omega \frac{\tilde{B}^{\ell}_{x,j}(\eta)}{2k\ell n}[\eta(z_{x,y})-\eta(z_{x,y}+\ell_i)][f(\eta)-f(\eta^{z_{x,y},z_{x,y}+\ell_i})]d\nu_b\Big|.\tag{56}$$

Our goal is to exchange the particles in the bond $\{z_{x,y},z_{x,y}+\ell_i\}$. In a similar way as it is done in Lemma 6.5 of [6], For every $x\in\mathbb{Z}$, $y\in\{1,\dots,\ell\}$ and every $i\in\{0,\dots,m-1\}$, we denote

$$\Omega_i(x,y) := \Big\{\eta \in \Omega : \overrightarrow{\eta}^{\ell}(z_{x,y}) \geq \frac{m}{\ell}\Big\} \cup \Big\{\eta \in \Omega : \overrightarrow{\eta}^{\ell}(z_{x,y}+\ell_i) \geq \frac{m}{\ell}\Big\}.$$

Thus, (56) is bounded from above by

$$\Big|\sum_x \Phi_n(s,\tfrac{x}{n}) \sum_{i=0}^{k-1}\sum_{y=1}^{\ell} \int_{\Omega-\Omega_i(x,y)} \frac{\tilde{B}^{\ell}_{x,j}(\eta)}{2k\ell n}[\eta(z_{x,y})-\eta(z_{x,y}+\ell_i)]\times \\ \times [f(\eta)-f(\eta^{z_{x,y},z_{x,y}+\ell_i})]\,d\nu_b\Big| \tag{57}$$

$$+\Big|\sum_x \Phi_n(s,\tfrac{x}{n}) \sum_{i=0}^{k-1}\sum_{y=1}^{\ell} \int_{\Omega_i(x,y)} \frac{\tilde{B}^{\ell}_{x,j}(\eta)}{2k\ell n}[\eta(z_{x,y})-\eta(z_{x,y}+\ell_i)]\times \\ \times [f(\eta)-f(\eta^{z_{x,y},z_{x,y}+\ell_i})]\,d\nu_b\Big|. \tag{58}$$

We note that for $\eta \in \Omega-\Omega_i(x,y)$, we have $\eta(z_{x,y})=\eta(z_{x,y}+\ell_i)=0$ for at least $\ell-2m+2$ values of $y\in\{1,\dots,\ell\}$. Plugging this with $\frac{1}{n}\sum_x|\Phi_n(s,\frac{x}{n})|\leq M_1$, $|\eta(\cdot)|\leq 1$ (which leads to $|\tilde{B}^{\ell}_{x,j}(\eta)|\leq 1$) and f being a density with respect to ν_b, we get that (57) is bounded from above by a constant times $1/\ell$. Due to our choice of ℓ and since $\gamma>0$, (57) vanishes as $n\to\infty$.

It remains to deal with (58), where we want to go from $\eta_{0,x,i,y} := \eta$ to $\eta^{z_{x,y},z_{x,y}+\ell_i}$. If $\overrightarrow{\eta}^{\ell}(z_{x,y})\geq m/\ell$, the strategy is the following: for any configuration $\eta\in\Omega_i(x,y)$, denote by $x_1,x_2,\dots,x_m$ the positions of the m particles inside the box $\{z_{x,y}+1,\dots,z_{x,y}+\ell\}$ closest to $z_{x,y}$. With at most $m\ell$ nearest-neighbor jumps, we can move the particles at $x_1,x_2,\dots,x_m$ to $z_{x,y}+1, z_{x,y}+2,\dots,z_{x,y}+m$, respectively.

On the other hand, if $\eta\in\Omega_i(x,y)$ and $\overrightarrow{\eta}^{\ell}(z_{x,y})<m/\ell$ then necessarily we have $\overrightarrow{\eta}^{\ell}(z_{x,y}+\ell_i)\geq m/\ell$; in this case, denote by $x_1,x_2,\dots,x_m$ the positions of the m particles inside the box $\{z_{x,y}+\ell_i+1,\dots,z_{x,y}+\ell_i+\ell\}$ closest to $z_{x,y}$. With at most $m\ell$ nearest-neighbor jumps, we can move the particles at $x_1,x_2,\dots,x_m$ to $z_{x,y}+\ell_i+1, z_{x,y}+\ell_i+2,\dots,z_{x,y}+\ell_i+m$, respectively.

In both cases, we denote the configuration with the group of at least m particles in consecutive sites next to $z_{x,y}$ (resp. $z_{x,y}+\ell_i$) by $\eta_{1,x,i,y}$. Then, we exchange the particles in the bond $\{z_{x,y},z_{x,y}+\ell_i\}$ by applying Lemma 22. At this point, our configuration is $\eta_{2,x,i,y} := (\eta_{1,x,i,y})^{z_x+y,z_x+y+\ell_i}$. Finally, we use nearest-neighbor jumps in order to bring the m auxiliary particles back to their initial positions $x_1,x_2,\dots,x_m$. We observe that our configuration now is exactly $\eta_{3,x,i,y} := \eta^{z_{x,y},z_{x,y}+\ell_i}$. In this way, (58) is bounded from above by the sum of

$$\Big|\sum_x \Phi_n(s,\tfrac{x}{n}) \sum_{i=0}^{k-1}\sum_{y=1}^{\ell} \int_{\Omega_i(x,y)} \frac{\tilde{B}^{\ell}_{x,j}(\eta)}{2k\ell n}[\eta(z_{x,y})-\eta(z_{x,y}+\ell_i)] \\ \times\big([f(\eta)-f(\eta_{1,x,i,y})]+[f(\eta_{2,x,i,y})-f(\eta_{3,x,i,y})]\big)\,d\nu_b\Big|, \tag{59}$$

$$+\Big|\sum_x \Phi_n(s,\tfrac{x}{n})\sum_{i=0}^{k-1}\sum_{y=1}^{\ell}\int_{\Omega_i(x,y)}\frac{\tilde{B}^{\ell}_{x,j}(\eta)}{2k\ell n}[\eta(z_{x,y})-\eta(z_{x,y}+\ell_i)] \\ \times[f(\eta_{1,x,i,y})-f(\eta_{2,x,i,y})]\,d\nu_b\Big|, \tag{60}$$

In order to treat (59), we write

$$[f(\eta)-f(\eta_{1,x,i,y})]+[f(\eta_{2,x,i,y})-f(\eta_{3,x,i,y})]=\sum_{r\in I^{NN}_{x,i,y}}[f(\eta^{(r-1)})-f(\eta^{(r)})],$$

where (for every fixed x, i, y) $I^{NN}_{x,i,y}$ is the set of bonds in which we use nearest-neighbor jumps. From Young's inequality and (25), (59) is bounded from above by a constant times

$$\frac{\ell M_1}{A}+\frac{AM_2}{k\ell n}\sum_{i=0}^{k-1}\int_{\Omega}\sum_x\sum_{y=1}^{\ell}\sum_{r\in I^{NN}_{x,i,y}}\left[\sqrt{f(\eta^{(r-1)})}-\sqrt{f(\eta^{(r)})}\right]^2 d\nu_b,$$

for any $A>0$. Above we used the fact that the number of bonds of $I^{NN}_{x,i,y}$ is bounded by $C\ell$ (where C is a positive constant independent of x, i and y), $|\eta(\cdot)|\leq 1$ and f is a density with respect to ν_b. We observe that in the triple summation inside the integral over Ω above (here we fix $i\in\{0,\ldots,k-1\}$), the number of times that a fixed bond $\{z_0-1,z_0\}$ is counted is at most of order ℓ^2. Thus, From (40), (59) is bounded from above by a constant times

$$\frac{\ell M_1}{A}+\frac{AM_2}{n}\ell\mathfrak{D}(\sqrt{f},\nu_b), \tag{61}$$

for any $A>0$. Next, we treat (60), which deals mostly with long jumps. With a similar reasoning as we did with (59), (60) is bounded from above by

$$\frac{M_1}{\tilde{A}}+\frac{\tilde{A}M_2}{k\ell n}\sum_{i=0}^{k-1}\sum_{y=1}^{\ell}\sum_x\int_{\Omega_i(x)}\left[\sqrt{f(\eta_{1,x,i,y})}-\sqrt{f\big((\eta_{1,x,i,y})^{z_x+y,z_x+y+\ell_i}\big)}\right]^2 d\nu_b,$$

for every $\tilde{A}>0$. Above we made use of (25). Applying Lemma 22, we conclude that (60) is bounded from above by a constant times

$$\frac{M_1}{\tilde{A}}+\frac{\tilde{A}M_2}{k\ell n}\sum_{i=0}^{k-1}\sum_{y=1}^{\ell}(\ell_i)^{\gamma}\mathfrak{D}(\sqrt{f},\nu_b)\leq\frac{M_1}{\tilde{A}}+\tilde{A}M_2\varepsilon^{\gamma}m^{\gamma}n^{\gamma-1}\mathfrak{D}(\sqrt{f},\nu_b), \tag{62}$$

for any $\tilde{A}>0$. Above we used the fact that $\ell_i\leq mk\ell=m\varepsilon n$. Since $\ell=\varepsilon n^{\gamma/2}$, by choosing $A=n^{\gamma}\varepsilon^{\gamma/2}(2M_2\ell)^{-1}$ in (61) and $\tilde{A}=\varepsilon^{-\gamma/2}(2M_2m^{\gamma})^{-1}$ in (62), from (42) we conclude that the display in (54) is bounded from above by a constant times

$$M_1 M_2 \ell^2 \varepsilon^{-\gamma/2} n^{-\gamma} + M_1 M_2 \varepsilon^{\gamma/2} = M_1 M_2 (\varepsilon^{2-\frac{\gamma}{2}} + \varepsilon^{\gamma/2}),$$

which vanishes for any $\gamma \in (0, 2)$ when $\varepsilon \to 0^+$, ending the proof. □

Acknowledgements P.C. was funded by the Deutsche Forschungsgemeinschaft (DFG, German Research Foundation) under Germany's Excellence Strategy—EXC 2047/1—390685813.

P.G. thanks FCT/Portugal for financial support through the projects UIDB/04459/2020 and UIDP/04459/2020. This project has received funding from the European Research Council (ERC) under the European Union's Horizon 2020 research and innovative programme (grant agreement n. 715734).

References

1. Baldasso, R., Menezes, O., Neumann, A., Souza, R.R.: Exclusion process with slow boundary. J. Stat. Phys. **167**(5), 1112–1142 (2017). https://doi.org/10.1007/s10955-017-1763-5
2. Bernardin, C., Cardoso, P., Gonçalves, P., Scotta, S.: Hydrodynamic limit for a boundary driven super-diffusive symmetric exclusion. Stoch. Process. Appl. **165**, 43–95 (2023). https://doi.org/10.1016/j.spa.2023.08.002
3. Bernardin, C., Gonçalves, P., Jiménez-Oviedo, B.: A microscopic model for a one parameter class of fractional Laplacians with Dirichlet boundary conditions. Arch. Ration. Mech. Anal. **239**(1), 1–48 (2021). https://doi.org/10.1007/s00205-020-01549-9
4. Billingsley, P.: Convergence of probability measures, second edn. Wiley Series in Probability and Statistics: Probability and Statistics. Wiley, New York (1999). https://doi.org/10.1002/9780470316962. A Wiley-Interscience Publication
5. Bonorino, L., De Paula, R., Gonçalves, P., Neumann, A.: Hydrodynamics for the porous medium model with slow reservoirs. J. Stat. Phys. **179**(3), 748–788 (2020). https://doi.org/10.1007/s10955-020-02550-y
6. Cardoso, P., De Paula, R., Gonçalves, P.: Derivation of the fractional porous medium equation from a microscopic dynamics. Nonlinearity **36**(3), 1840–1872 (2023). https://doi.org/10.1088/1361-6544/acb7c1
7. Cardoso, P., Gonçalves, P., Jiménez-Oviedo, B.: Hydrodynamic behavior of long-range symmetric exclusion with a slow barrier: diffusive regime. Ann. Inst. Henri Poincaré Probab. Stat. **60**(3), 1511–1542 (2024). https://doi.org/10.1214/23-AIHP1365
8. Cardoso, P., Gonçalves, P., Jiménez-Oviedo, B.: Hydrodynamic behavior of long-range symmetric exclusion with a slow barrier: superdiffusive regime. Ann. Sc. Norm. Super. Pisa Cl. Sci. (5) **25**(4), 1813–1876 (2024). https://doi.org/10.2422/2036-2145.202203_019
9. De Pablo, A., Quirós, F., Rodríguez, A., Vázquez, J.L.: A fractional porous medium equation. Adv. Math. **226**(2), 1378–1409 (2011). https://doi.org/10.1016/j.aim.2010.07.017
10. De Pablo, A., Quirós, F., Rodríguez, A., Vázquez, J.L.: A general fractional porous medium equation. Comm. Pure Appl. Math. **65**(9), 1242–1284 (2012). https://doi.org/10.1002/cpa.21408
11. Gonçalves, P., Landim, C., Toninelli, C.: Hydrodynamic limit for a particle system with degenerate rates. Ann. Inst. Henri Poincaré Probab. Stat. **45**(4), 887–909 (2009). https://doi.org/10.1214/09-AIHP210
12. Gonçalves, P., Nahum, G., Simon, M.: From exclusion to slow and fast diffusion. To appear in Ann. Inst. Henri Poincaré Probab. Stat (2023)
13. Guo, M.Z., Papanicolaou, G.C., Varadhan, S.R.S.: Nonlinear diffusion limit for a system with nearest neighbor interactions. Comm. Math. Phys. **118**(1), 31–59 (1988). http://projecteuclid.org/euclid.cmp/1104161907

14. Jara, M.: Hydrodynamic limit of particle systems with long jumps. `arXiv: /0805.1326` (2009). https://arxiv.org/abs/0805.1326
15. Kipnis, C., Landim, C.: Scaling limits of interacting particle systems, Grundlehren der mathematischen Wissenschaften [Fundamental Principles of Mathematical Sciences], vol. 320. Springer, Berlin (1999). https://doi.org/10.1007/978-3-662-03752-2
16. Spitzer, F.: Interaction of Markov processes. Adv. Math. **5**, 246–290 (1970). https://doi.org/10.1016/0001-8708(70)90034-4

On Bi-Lipschitz Classification of Real and Complex Singularities

Alexandre Fernandes

Abstract Lipschitz geometry of singularities is the local study of algebraic/analytic (semi-algebraic/subanalytic) sets up to bi-Lipschitz homeomorphisms. Results by Mostowski [35] and Parusiński [40] guarantee that there is only infinite many countable of such singularities up to bi-Lipschitz homeomorphism, in other words, they can be classified by means of discrete invariants. This text aims to introduce the known results of the bi-Lipschitz classification of singularities; these notes complement and intersect the articles that also provide an introduction to the subject [20, 28, 39, 43, 44].

1 Introduction

In 1985, T. Mostowski proved the following Lipschitz finiteness theorem: any complex algebraic family of subsets of $\mathbb{C}^n$ has finitely many representatives up to bi-Lipschitz (with respect to the outer metric) homeomorphisms. In 1997, A. Parusinski gave the following real version of the Mostowski's result: any subanalytic family of subsets of $\mathbb{R}^n$ has finitely many representatives up to bi-Lipschitz (with respect to the outer metric) homeomorphisms. It is important to mention that those results above say nothing about how the sets can be classified up to bi-Lipschitz homemomorphisms; roughly speaking, those results just say that there is a way to use discrete invariants to classify those sets up to bi-Lipschitz homeomorphisms. In that direction, in 1999, L. Birbrair presented one of the first classification of singulatrities up bi-Lipschitz homeomorphisms.

Theorem of Birbrair. *Let (X, x_0) be a 2D real singularity which is homeomorphic to a real disc. If (X, x_0) is equipped with the inner metric, then there is a rational number $\beta \geq 1$ such that (X, x_0) is bi-Lipschitz homeomorphic to the germ, at origin $\mathbf{0} \in \mathbb{R}^3$ of the β-horn*

$$H_\beta: \ x^2 + y^2 = z^{2\beta} \quad z \geq 0.$$

A. Fernandes (✉)
Federal University of Ceará, Fortaleza, CE, Brazil
e-mail: alex@mat.ufc.br

M. Abreu et al. (eds.), *Brazil-Portugal Mathematics*, Coimbra Mathematical Texts 4,
https://doi.org/10.1007/978-3-031-98455-6_5

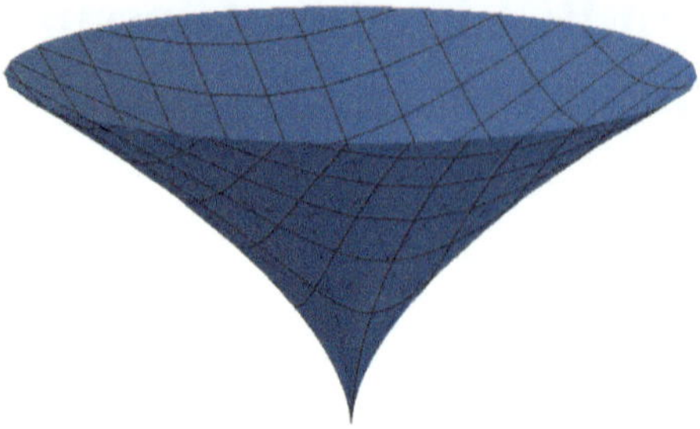

Fig. 1 A β-horn H_β with $\beta > 1$

Moreover, two of those horns $(H_\beta, 0)$ *and* $(H_{\beta'}, 0)$ *are bi-Lipschitz homeomorphic iff* $\beta = \beta'$ (Fig. 1).

Theorem 14 demonstrated the Lipschitz geometry of real singularities can be very rich even in low dimensions. In contrast to that, as an easy consequence of the same result above, any complex curve singularity (equipped with the inner metric) is bi-Lipschitz homeomorphic to the cone of its link. In other words, for complex curve singularities, the inner Lipschitz geometry is nothing else than topology of the singularities. Since then, the problem of classifying singularities up to bi-Lipschitz homeomorphisms has been intensively developed.

This article aims to present the Lipschitz geometry of singularities and some of the main results in the area for real and complex singularities in low dimensions. The results are presented without their respective proofs, but they are followed by their appropriate bibliographic references.

In the following, we describe the sections of this article.

- Section 2 introduces the notion of singularities (real anda complex) we are interested in. We end this section by introducing the notion of diffeomorphic and homeomorphic singularities, and through an example from Hassler Whitney and the Local Conic Structure Theorem, we show why the classification of singularities by diffeomorphisms seems much more rigid than by homeomorphism ones.
- Section 3, by a precise definition of bi-Lipschitz homeomorphism of singularities, explains exactly what we mean by Lipschitz Geometry of Singularities. In this section, we also give an interpretation of the Lipschitz Fineteness Theorems of Tadeusz Mostowski and Adam Parusinski.
- Section 4 address the Theorem of Lev Birbrair which completely solves the classification problem of 2d real singularities up to inner bi-Lipschitz homeomorphisms. Section 4 also introduces a complete invariant to the outer bi-Lipschitz classification of 2D real singularities of the type $T_f \cap G_f$ where $f : (T_f, 0) \to \mathbb{R}$ is the germ of a Lipschitz subanalytic nonnegative function and T_f is a subanalytic topological triangle in $\mathbb{R}^2$ with vertex at 0. This invariant is the so-called *pizza of* f.
- One starts Sect. 5 with recently proved theorems that show that germs of complex analytic sets which are bi-Lipschitz homeomorphic to smooth germs must be smooth. Section 5 also presents results with complete answers on the bi-Lipschitz classification of complex singularities. The section presents the outer bi-Lipschitz

classification of singularities of plane algebraic curves that was pioneered by B. Teissier and F. Pham in 1969 (see [41, 42]), the section still shows the developments of this Pham-Teissier result to the present day. Regarding complex surface singularities, Sect. 5 also presents the first inner bi-Lipschitz invariants, so-called "fast loops" and "separating sets", which were used in the seminal article [6] to show that there are singularities of complex surfaces that are not inner bi-Lipschitz homeomorphic to right cones. Finally, this section states the inner bi-Lipschitz classification normal complex surface singularities due to L. Birbrair, W. Neumann and A. Pichon (see [19]).

- Section 6 is the last section of this article, it shows two global bi-Lipschitz classification results, the first one is the outer bi-Lipschitz classification of plane complex curves, it is a theorem of R. Targino published in [47]. The second one is the inner bi-Lipschitz classification of real 2D semi-algebraic closed subsets in $\mathbb{R}^n$, is a theorem published in the joint work with Sampaio [30]. This section ends with two open problems on the subject.

2 Singular Points

Let us start this section explaining what we mean by singularities of a subset $X \subset \mathbb{K}^n$ (where $\mathbb{K} = \mathbb{R}$ or $\mathbb{C}$).

Definition 1 A point $p \in X$ is called a *smooth point* of X if there exists an open neighborhood U of p in X which is diffeomorphic to an open ball $B \subset \mathbb{K}^d$, in this case p is a smooth point of *dimension* d.

Let $X \subset \mathbb{K}^n$ be an analytic subset. It is well known that the subset of points in X which are not smooth is nowhere dense; actually, it is a null measure subset. This observation motivates us to call smooth points of X by *regular* and to give the following special name for the points of X which are not regular.

Definition 2 A point $p \in X$ is called a *singular point* of X if it is not a smooth point of X.

Roughly speaking, the study of the geometry of singularities is the research field in Mathematics in charge to describe as precisely as possible what is the shape of singularities.

Before starting with the first examples, let us introduce a notion of tangent space of a set X at a point $p \in X$.

Definition 3 Given a point $p \in X \subset \mathbb{K}^n$, the set

$$T_pX = \{v \in \mathbb{K}^n \ : \exists\, \gamma : (-\epsilon, \epsilon) \to X,\ \gamma(0) = p \text{ and } v = \gamma'(0)\}$$

is called the *tangent space of X at p*.

Proposition 4 *If p is a d-dimensional smooth point of $X \subset \mathbb{K}^n$, then $T_q X$ is a d-dimensional $\mathbb{K}$ vector subspace of $\mathbb{K}^n$ for any q in a neighborhood of p in X.*

Among other properties the tangent space $T_p X$ is usefull to recognize singular points of X as we can see in the next example.

Example 5 Let $X \subset \mathbb{K}^2$ be the cusp $X = \{(z, w) \in \mathbb{K}^2 : z^2 = w^3\}$. We claim that $\mathbf{0} = (0, 0)$ is a singular point of X. Indeed, it follows from the Implicit Function Theorem that $T_q X$ is a $\mathbb{K}$ vector subspace of $\mathbb{K}$ of dimension 1 for any $\nvdash \neq q \in X$. On the other hand, we can see that $T_{\mathbf{0}} X$ is the null vector subspace of $\mathbb{K}^2$. By Proposition 4, $\mathbf{0}$ is a singular point of X.

Given a point $p \in X \subset \mathbb{K}^n$, the pair (X, p) is called the *germ of X at p*. To give some sense to the main question of this section we introduce the following definitions.

Definition 6 Two germs (X, x_0) and (Y, y_0) are said *diffeomorphic* if there exists neighborhoods $U \subset X$ of x_0 and $V \subset Y$ of y_0 which are diffeomorphic.

In 1965, with the following simple yet fundamental example, H. Whitney showed that is not possible to describe all the germs of analytic singularities up to diffeomorphims just by using discrete invariants.

Example 7 (*Whitney 1965*) Let

$$X_t = \{(z, w) \in \mathbb{K}^2 : z \cdot w \cdot (z - w) \cdot (z - tw) = 0.\}$$

For generic $s \neq t$, H. Whitney noted that $(X_s, \mathbf{0})$ and $(X_t, \mathbf{0})$ are not diffeomorphic.

Roughly speaking, Whitney's example gave us a family of singularities that look similar to the naked eye but that are distinct, when you look them up to diffeomorphisms. Somehow this example motivates the following definition that allows us to study singularities up to geometric transformations which are less rigid than diffeomorphisms.

Definition 8 Two germs (X, x_0) and (Y, y_0) are said *homeomorphic* if there exists neighborhoods $U \subset X$ of x_0 and $V \subset Y$ of y_0 which are homeomorphic.

In the context of germ of homeomorphisms, we have many natural singularities that are not homeomorphic to each other. The natural way to approach such examples is by making use of the following structural result.

Theorem 9 (Local Conical Structure) *Let $X \subset \mathbb{K}^n$ be a subanalytic subset and let $p \in X$. There exists a small $\epsilon_0 > 0$ such that:*

1. *for each $0 < \epsilon < \epsilon_0$, X is transversal to the spheres $\mathbb{S}(p, \epsilon)$, centered at p with radius ϵ;*

2. *for each* $0 < \epsilon < \epsilon_0$*, there is a homeomorphism* ϕ *from* $X \cap B(p, \epsilon)$ *onto the real cone over* $X \cap \mathbb{S}(p, \epsilon)$*, where* B *is the Euclidean ball centered at* p *with radius* ϵ*, such that*

$$|\phi(x) - p| = |x - p| \; \forall \, x \in X \cap B(p, \epsilon).$$

For $\epsilon > 0$ smaller than ϵ_0 given as in the Local Conic Structure Theorem, the intersection $X(\epsilon) = X \cap \mathbb{S}(p, \epsilon)$ is called the *link of* (X, p).

3 Lipschitz Geometry

Let us start this section recalling the definition of Lipschitz maps between two metric spaces.

Definition 10 Let A and B metric spaces with the respective metrics d_A and d_B. A mapping $F: A \to B$ is said *Lipschitz* (with constant $\lambda > 0$) if:

$$d_B(F(x), F(y)) \leq \lambda d_A(x, y) \; \forall \, x, y \in A.$$

If $F: A \to B$ exists and it is also a Lipschitz mapping, F is said a *bi-Lipschitz mapping*. In this case, A and B are are said *bi-Lipschitz homeomorphic*.

Unless otherwise mentioned, we consider the subsets $X \subset \mathbb{K}^n$ equipped with the Euclidean metric: $d_{out}(x, y) = \|x - y\|, \forall \, x, y \in X$ which we call the *outer metric of* X.

Definition 11 Two germs of Euclidean subsets (X, x_0) and (Y, y_0) are said *bi-Lipschitz homeomorphic* if there exists neighborhoods $U \subset X$ of x_0 and $V \subset Y$ of y_0 which are bi-Lipschitz homeomorphic.

We are able to say what we mean by **Lipschitz Geometry of Singularities**, namely: it is the research field in Mathematics in charge of classifying, listing models and recognizing germs (X, p), with $X \subset \mathbb{C}^n$ algebraic or analytic, with $X \subset \mathbb{R}^n$ semialgebraic or subanalytic, up to bi-Lipschitz homeomorphisms.

Theorem 12 (Bi-Lipschitz Finiteness Theorem)

- **(Mostowski** *[35]***)** *For algebraic subsets in* $\mathbb{C}^n$*, there are only infinite many countable germs* (X, p) *up to bi-Lipschitz homeomorphisms.*
- **(Parusiński** *[40]***)** *For subnalytic subsets in* $\mathbb{R}^n$*, there are only infinite many countable pairs* (X, p) *up to bi-Lipschitz homeomorphisms.*

The bi-Lipschitz Fineteness Theorem is one of the fundamental results that support the Lipschitz Geometry of Singularities; roughly speaking, this theorem says that it is possible to classify those "singularities" by using discrete invariants, however, the theorem and its respective proof say nothing about how to obtain such classifications. As we already mentioned, the main goal os this notes is to present a list of classification theorems and some open questions in Lipschitz Geometry of Singularities.

4 Lipschitz Geometry of Real Singularities

Let $X \subset \mathbb{R}^n$ be a connected subanalytic subset. Given $x, y \in X$, let Γ_{xy} be the set of paths γ on X connecting x to y. It is known that the function $(x, y) \in X \times X \mapsto d_{inn}(x, y)$, where

$$d_{inn}(x, y) = \inf\{length(\gamma) \ : \ \gamma \in \Gamma_{x,y}\}$$

defines a metric on X. It is called the *inner metric on* X.

Given a germ of subanalytic set with connected link (X, p), let us denote by $(X, p)_{inn}$ the germ equipped with the inner metric on X. So, the following result says that $(X, p)_{inn}$ is an outer bi-Lipschitz invariant of (X, p). Since $(X, p)_{inn}$ is far from being a complete invariant, it is not a so good result toward the bi-Lipschitz classification of singularities, but it is something!

Proposition 13 *Let (X, x_0) and (Y, y_0) be germs of subanalytic sets with connected links. If (X, x_0) and (Y, y_0) are bi-Lipschitz homeomorphic (with respect to the outer metric), then $(X, x_0)_{inn}$ and $(Y, y_0)_{inn}$ are bi-Lipschitz homeomorphic (with respect to the inner metric).*

In the seminal paper [2], published in 1999, Lev Birbrair presented one of the first classification of singulatrities up bi-Lipschitz homeomorphisms.

Theorem 14 (Birbrair [2]) *If (X, x_0) is a germ of a 2-dimensional real subanalytic set which is homeomorphic to a real disc, then there is a rational number $\beta \geq 1$ such that $(X, x_0)_{inn}$ is bi-Lipschitz homeomorphic to the germ, at origin $\mathbf{0} \in \mathbb{R}^3$, of the β-horn*

$$H_\beta\colon \ x^2 + y^2 = z^{2\beta} \quad z \geq 0.$$

Moreover, two of those horns $(H_\beta, 0)$ and $(H_{\beta'}, 0)$ are bi-Lipschitz homeomorphic iff $\beta = \beta'$.

The following definition was first introduced by Birbrair and Mostowski in the paper [18].

Definition 15 The germ of a subanalytic set (X, p) is called *Lipschitz normally embedded* if there is a real constant $c \geq 1$ such that $d_{out} \leq c \cdot d_{inn}$, where d_{out} and $d_{inn]}$ are the outer and inner metric, respectively, on X.

Since the publication of the paper [18], the notion of Lipschitz normally embedded sets has been intensively explored as can be seen in the references [1, 10, 17, 24, 26, 32–34, 36, 37, 46].

Example 16 By using that the β-horn is a revolution surface with profile curve $\{(x, y) \in \mathbb{R}^2\ y = x^\beta \text{ and } 0 \leq x \leq 1\}$, one can see that the germ of β-horns at $\mathbf{0} \in \mathbb{R}^3$ are Lipschitz normally embedded.

It is clear that Theorem 14 also gives the outer bi-Lipschitz classification of real 2-dimensional real subanalytic sets which are Lipschitz normally embedded. Next example shows that the classification of real surface singularities up to inner bi-Lipschitz homeomorphism is weaker than the classification of those singularities up to outer bi-Lipschitz homeomorphisms.

Example 17 Let $X = \{(x, y, z) \in \mathbb{R}^3 : x^2 + y^4 = z^6 \text{ with } z \geq 0\}$. By the Theorem of Birbrair, $(X, \mathbf{0})$ is bi-Lipschitz homoeomorphic to the germ of some β-horn at the origin with respect to the inner metric (actually, one can see that this β is 3/2), but they are not bi-Lipschitz homeomorphic with respect to the outer metric. Indeed, one can see that $\gamma_1(t) = (t^3, 0, t)$ and $\gamma_2(t) = (-t^3, 0, t)$ are arcs on X such that:

$$d_{out}(\gamma_1(t), \gamma_2(t)) \approx t^3 \text{ and } d_{inn}(\gamma_1(t), \gamma_2(t)) \approx t^{3/2} \text{ as } t \to 0^+,$$

hence the germ $(X, \mathbf{0})$ is not Lipschitz normally embedded. In particular, $(X, \mathbf{0})$ is not outer bi-Lipschitz homeomorphic to any β-horn.

Although the Theorem of Birbrair brings a complete answer for the classification of 2-dimensional real singularities up to bi-Lipschitz homeomorphisms with respect to the inner metric, the respective problem of classification with respect to the outer metric remains open until now.

The two articles listed below correspond to the first attempts to obtain partial answers to the problem of bi-Lipschitz classification of 2-dimensional real germs with respect to the outer metric.

- In the paper [8], we completely solved the classification problem of 2-dimensional real subanalytic germs $(X, \mathbf{0})$ with $X = \Gamma_1 \cup \Gamma_2$ where Γ_1 and Γ_2 are topological triangles such that $\Gamma_1 \subset \mathbb{R}^2$ and Γ_2 is the graph of a subanalytic Lipschitz function germ $f : \Gamma_1 \to \mathbb{R}$.
- In the recent publication [15], Birbrair and Gabrielov extended the result above in the sense that they completely classified 2-dimensional real subanalytic germs $(X, \mathbf{0})$ with $X = \Gamma_1 \cup \Gamma_2$ where Γ_1 and Γ_2 are topological triangles such that Γ_1 and Γ_2 are topological triangles where their respective metrics outer and inner are comparable, in other words, they are Lipschitz normally embedded, i.e. ($\exists$ constant $c \geq 1$ s.t. $d_{inn} \leq c \cdot d_{out}$).

In the following, we briefly describe what is done in the paper [8].

Let Γ_1 and Γ_2 be germs of semianalytic arcs at the origin $\mathbf{0} \in \mathbb{R}^2$. It is clear that $\Gamma_1 \cup \Gamma_2$ separates the germ $(\mathbb{R}^2, \mathbf{0})$ into two connected components; let us say T_1 and

T_2, the component which has smaller density at $\mathbf{0}$ is said the *Hölder triangle* with boundary arcs Γ_1 and Γ_2, where the density of T_1 and T_2 at $\mathbf{0}$ are the real numbers:

$$\text{density}(T_j, \mathbf{0}) = \lim_{r \to 0+} \frac{1}{r^2} \text{area}(T_j \cap (B(\mathbf{0}, r))) \quad j = 1, 2,$$

where $B(\mathbf{0}, r) = \{z \in \mathbb{R}^2 \ : \ \|z\| \leq r\}$.

Let $f \colon (T_f, \mathbf{0}) \to (\mathbb{R}, 0)$ be a germ of nonnegative Lipschitz and subanalytic function defined on the germ of the Hölder triangle T_f at the origin $\mathbf{0} \in \mathbb{R}^2$. Let us denote by $(G_f, \mathbf{0})$ the germ of the graph of f at $\mathbf{0} \in \mathbb{R}^3$ and denote $X_f = G_f \cup T_f$.

Definition 18 ([5]) Let $f, g \colon (T_f, \mathbf{0}) \to (\mathbb{R}, 0)$ be nonegative Lipschitz subanalytic function germs define on Hölder triangles at $\mathbf{0} \in \mathbb{R}^2$. The germs f and g are said *Lipschitz contact equivalent* if, and only if, $(X_f, \mathbf{0})$ and $(X_g, \mathbf{0})$ are bi-Lipschitz homeomorphic.

Lipschitz contact. Let Γ_1 and Γ_2 be germs of subanalytic arcs at $\mathbf{0} \in \mathbb{R}^n$ (germ of subanalytic set which the link is only one point) and let γ_1 and γ_2 the respective parametrizations of Γ_1 and Γ_2 by the distance to the origin $\mathbf{0} \in \mathbb{R}^n$. The rational number below is said the *Lipschitz contact* between Γ_1 and Γ_2:

$$cont(\Gamma_1, \Gamma_2) = \lim_{t \to 0^+} \frac{log(|\gamma_1(t) - \gamma_2(t)|)}{log(t)}.$$

Given a Hölder triangle T at $\mathbf{0} \in \mathbb{R}^2$, with boundary arcs Γ_1 and Γ_2, let us denote $exp(T) := cont(\gamma_1, \gamma_2)$.

Order of f along an arc. Given a nonnegative subanalytic function germ

$$f \colon (T_f, \mathbf{0}) \to (\mathbb{R}, 0),$$

for each subanalytic arc $\Gamma \subset T_f$ at $\mathbf{0}$, we define the *order of f along* Γ by:

$$ord_\Gamma f = \lim_{t \to 0^+} \frac{log(f(\gamma(t)))}{log(t)}$$

where γ is a parametrization of Γ such that $|\gamma(t)| \sim t$.

Example 19 Let $T = \{(x, y) \in \mathbb{R}^2; \ x \geq 0, \ y \geq 0\}$. Let f and g be the function germs defined on $(T, \mathbf{0})$ by $f(x, y) = x^4 + y^2$ and $g(x, y) = x^4 + xy + y^2$. Let us see that f and g are not Lipschitz contact equivalent. In fact, let us suppose, by contradiction, that ϕ is a germ of bi-Lipschitz homeomorphism $\phi \colon (T, 0) \to (T, 0)$ which makes f and g Lipschitz contact equivalent. In this case, we have the boundary arcs of T are invariant by ϕ. Let Γ be the horizontal boundary arc of T. So, we have the order of f along any arc which has contact $\alpha \geq 1$ with Γ is either 2α if $\alpha \leq 2$ or 4 if $\alpha \geq 2$. On the other hand, the order of g along any arc which has contact $\alpha \geq 1$

with Γ is either $\alpha + 1$ if $\alpha \leq 3$ or 4 if $\alpha \geq 3$. Since ϕ sent arcs with contact α with Γ to arcs with the same property, it follows that ϕ can not conjugates f and g in the terms of the Lipschitz contact equivalence.

We use the ideas of the example above to introduce the last ingredient of our invariant.

Width of f along an arc. Given a germ of a nonnegative subanalytic function $f\colon (T_f, \mathbf{0}) \to (\mathbb{R}, 0)$, for each subanalytic arc $\Gamma \subset T_f$, at $\mathbf{0}$, we define the *width of f along* Γ by:

$$\mu_{T_f}(f, \Gamma) = \inf = \{exp(T) \ : \ T \in \Lambda\}$$

where Λ is the family triangles $T \subset T_f$ such that $\Gamma \subset T$ and $ord_{\tilde{\Gamma}} f$ does not depend on $\tilde{\Gamma} \subset T$.

Theorem 20 (Birbrair-Fernandes-Gabrielov-Grandjean [8]) *Let $f\colon (T, \mathbf{0}) \to (\mathbb{R}, 0)$ be a nonnegative subanalytic continuous function germ. There exists a finite partition $\{T_i\}$ by Hölder triangles $T_i \subset T$ such that:*

1. *The set $Q_i = \{ord_\Gamma f \ : \ \Gamma \subset T_i\}$ is a closed interval in $\mathbb{Q} \cup \{\infty\}$. We denote $Q_i = [a_i, b_i]$ with $a_i < b_i$ or $b_i < a_i$ or $a_i = b_i$.*
2. *The function $\mu_i\colon Q_i \to \mathbb{Q} \cup \{\infty\}$ given by $\mu_i(q) = \mu_{T_i}(f, \Gamma)$ where Γ is any arc in T_i such that $ord_\Gamma f = q$ is well defined. Moreover, μ_i is an affine function with rational coefficients.*

The finite set $\{exp(T_i), Q_i, \mu_i\}$ described in Theorem 20 is said *a pizza of* $f\colon (T_f) \to (\mathbb{R}, 0)$. Of course, we can not expect uniqueness of pizzas of f, for instance, by taking subdivisions of the triangles T_{is} of a given pizza $\{exp(T_i), Q_i, \mu_i\}$ we get another pizza of f with more "pieces" than the original one. A pizza of f with minimal number of triangles is said *a minimal pizza of f*.

A finite family $\{\beta_i, Q_i, \mu_i\}$ is said *an abstract pizza* if:

1. each β_i is a rational number greater than or equal to 1;
2. each $Q_i = [a_i, b_i]$ is a closed interval in $\mathbb{Q} \cup \{\infty\}$ with $a_i < b_i$ or $b_i < a_i$ or $a_i = b_i$;
3. each $\mu_i\colon Q_i \to \mathbb{Q} \cup \{\infty\}$ is an affine function with rational coefficients and such that $\mu_i \geq \beta_i$.

Let us denote by H_f the abstract pizza provided by a minimal pizza of f.

Example 21 Let $T = \{(x, y) \in \mathbb{R}^2;\ x \geq 0,\ y \geq 0\}$. Let $f\colon (T, \mathbf{0}) \to (\mathbb{R}, 0)$ given by $f(x, y) = |x^2 - y^3|$. In this case, H_f is the following abstract pizza with two pieces:

1. $\beta_1 = \dfrac{3}{2}$, $Q_1 = [3, \infty]$, $\mu_1(q) = q - \dfrac{3}{2}$;
2. $\beta_2 = 1$, $Q_2 = [\infty, 2]$, $\mu_2(q) = \dfrac{q}{2}$.

Theorem 22 (Birbrair-Fernandes-Gabrielov-Grandjean [8]) *Let* $f\colon (T_f, \mathbf{0}) \to (\mathbb{R}, 0)$ *and* $g\colon (T_g, \mathbf{0}) \to (\mathbb{R}, 0)$ *be nonnegative subanalytic Lipschitz function germs. Then,* f *is Lipschitz contact equivalent to* g *if, and only if,* $H_f = H_g$.

We end this section by emphasizing that the problem of classifying 2-dimensional real subanalytic germs up to bi-Lipschitz homeomorphisms remains open. It is important to mention that Lev Birbrair and Andrei Gabrielov have published significant general results on the bi-Lipschitz classification of 2D real singularities that point in the direction of a possible complete invariant for this classification problem (see [15, 16]).

5 Lipschitz Geometry of Complex Singularities

One of the crudest question on bi-Lipschitz geometry of complex singularities was to recognize the germs of complex analytic sets which are bi-Lipschitz homeomorphic to the smooth ones, in other words, the germs which are not singular from the bi-Lipschitz point of view. Fortunataly, this question is completely solved as one can see in the next two theorems. In the paper [10], written jointly with L. Birbrair, E. Sampaio, and L. D. Trang, we obtained the first result towards the answers to this above question.

Theorem 23 (Birbrair-Fernandes-Lê-Sampaio [10]) *Let* $X \subset \mathbb{C}^n$ *be an analytic complex set. If* $x_0 \in X$ *is a point such that the germ* (X, x_0) *is subanalytically bi-Lipschitz homeomorphic to* $(\mathbb{C}^d, \mathbf{0})$*, then* x_0 *is a smooth point of* X.

In the paper [45], Sampaio improved the above theorem by showing that x_0 is a smooth point of X if there exist an outer bi-Lipschitz homeomorphism $\phi\colon (X, x_0) \to (\mathbb{C}^d, \mathbf{0})$ even without the subanalicity assumption on ϕ, more precisely he proved the following.

Theorem 24 (Sampaio [45]) *Let* $X \subset \mathbb{C}^n$ *be an analytic complex set. If* $x_0 \in X$ *is a point such that the germ* (X, x_0) *is bi-Lipschitz homeomorphic to* $(\mathbb{C}^d, \mathbf{0})$*, then* x_0 *is a smooth point of* X.

Before starting to describe what has been done on the bi-Lipschitz classification of complex singularities, let us recall the first significant results on the embedded topology of germs of complex algebraic plane curves were obtained in the beginning of 20th century. It can be seen with details in [23] (p. 442) that the embedded topology of such a germ is completely determined by the following combinatorial data; let us call it *Zariski data* of the germ:

- the sequence of Puiseux pairs of the branches of the germ;
- the intersection numbers of all pairs of branches of the germ.

In other words, the mentioned result that can be seen in [23] (p. 442) says that two germs of complex plane curves X and $\widetilde{X}$ are topologically equivalent, in the sense that there exists a germ of homeomorphism $(\mathbb{C}^2, \mathbf{0}) \to (\mathbb{C}^2, \mathbf{0})$ which sends X onto $\widetilde{X}$, if and only if, they have the same Zariski data.

Although some of the answers to the fundamental questions of Lipschitz geometry are recent, such as Theorems 12, 23 and 24, the very first result on the classification of complex singularities dates back to 1969, when Pham and Teissier was studying equisingularity in families of complex curves and they proved the following remarkable theorem (see [41, 42]).

Theorem 25 (Pham-Teissier [41, 42]) *Let X and $\widetilde{X}$ germs at $\mathbf{0}$ of complex plane curve. Then, the following conditions are equivalent:*

I. there exists a germ of a (meromorphic) bi-Lipschitz homeomorphism $F : X \to \widetilde{X}$;
II. X and $\widetilde{X}$ are topologically equivalent, as germs embedded in $\mathbb{C}^2$.

In 2003, this classification result was improved in the paper [25], where we replaced the adjective "meromorphic" of the bi-Lipschitz homeomorphism in Item I of Theorem 25 by the adjective "subanalytic". More precisely, the improvement of Theorem 25 was the following:

Theorem 26 (Fernandes [25]) *Let X and $\widetilde{X}$ germs at $\mathbf{0}$ of complex plane curve. Then, the following conditions are equivalent:*

I. there exists a germ of a subanalytic bi-Lipschitz homeomorphism $F : X \to \widetilde{X}$;
II. X and $\widetilde{X}$ are topologically equivalent, as germs embedded in $\mathbb{C}^2$.

Later on, in the paper [38] published in 2014, W. Neumann and A. Pichon were able to obtain the above theorem removing the adjective "meromorphic" and/or "subanalytic" of the bi-Lipschitz homeomorphism in Item I of the Theorem 25. Then, Theorem 25 got the following version:

Theorem 27 (Neumann-Pichon [38]) *Let X and $\widetilde{X}$ germs at $\mathbf{0}$ of complex plane curve. Then, the following conditions are equivalent:*

I. there exists a germ of a bi-Lipschitz homeomorphism $F : X \to \widetilde{X}$;
II. X and $\widetilde{X}$ are topologically equivalent, as germs embedded in $\mathbb{C}^2$.

Just to finish this sequence of results, let us mention the final improvement of it obtained by Fernandes, Sampaio and Silva in the paper [31] in terms of some "Hölder Geometry of Singularities" as we see below.

Theorem 28 (Fernandes-Sampaio-Silva [31]) *Let X and $\widetilde{X}$ be germs at $\mathbf{0}$ of complex plane curve. Then, the following conditions are equivalent:*

I. *For each* $0 < \alpha < 1$, *there exists a germ of a* α*-Hölder mapping* $F : X \to \widetilde{X}$ *with inverse* $F^{-1} : \widetilde{X} \to X$ *also* α*-Hölder;*
II. X *is topologically equivalent to* $\widetilde{X}$, *as germs embedded in* $\mathbb{C}^2$.

In higher dimensions, the problem of bi-Lipschitz classification of singularities has proved to be quite difficult. Actually, even in dimension 2, the problem of classifying the germs of complex analytic surfaces up to outer bi-Lipschitz homeomorphism still remains open. However, we already know good answers to the problem of classifying the germs of complex analytic surfaces up to inner bi-Lipschitz which were given by L. Birbrair, W. Neumann and A. Pichon; and, from now on, we start to explain what was obtained in that direction.

Let $X \subset \mathbb{C}^n$ be an analytic closed subset. Given a point $p \in X$, let us recall the ϵ-link of X at p, which we denote by $X_\epsilon(p)$, it is given by $X \cap \mathbb{S}(p, \epsilon)$ where $\mathbb{S}(p, \epsilon)$ is the Euclidean sphere centered at p with sufficiently small radius $\epsilon > o$.

Definition 29 A subanalytic subset $X \subset \mathbb{K}^n$ is said *inner Lipschitz conical at* $p \in X$ if: for some $\epsilon > 0$, $X \cap B(p, \epsilon)$ is inner bi-Lipschitz homeomorphic to the right cone over $X_\epsilon(p)$ with vertex at p.

The first examples of complex surface singularities that are not inner Lipschitz conical were obtained in 2005, together with L. Birbrair, we published these results in the paper [7]. In 2008, together with L. Brbrair and W. Neumann, we published a series of papers that, among other conclusions, showed that the complex surface singularities which are inner Lipschitz conical are rare (see [11–14]). Among those publications, we highlight [11, 14], where we present and develop the concept of **fast loops** and **separating sets**, respectively. Next, we describe the concepts of fast loops and separating sets and the main results we obtained about the functions of these objects as inner bi-Lipschitz invariant of singularities.

Fast loops. Given $\alpha > 0$, a Lipschitz map $\gamma : \mathbb{S}^1 \to X_\epsilon(x)$ is called an α*-fast loop* if there exists a homotopy H_t em $X \cap B(x, \epsilon)$; which deforms γ to x and satisfies the following volume growth condition:

$$\lim_{r \to 0^+} \frac{1}{r^a} \mathcal{H}^2(Im(H) \cap B(x, r)) = 0 \; \forall \; 0 < a < \alpha.$$

Where $\mathcal{H}^2$ denotes the 2-dimensional Hausdorff measure with respect to the Euclidean metric.

Theorem 30 (Birbrair-Fernandes-Neumann [11]) *Existence of fast loops which are homotopically nontrivial in the link is an inner bi-Lipschitz invariant of the germs of subanalytic sets. Moreover, if a germ with isolated singularity contains a homotopic nontrivial* α*-fast loop, with* $\alpha > 2$, *then it is not inner Lipschitz conical.*

Next example shows a germ of complex algebraic surface with isolated singularity which contains a homotopically nontrivial $\frac{5}{2}$-fast loop; in particular, it is not inner Lipschitz conical.

Example 31 (*Birbrair-Fernandes-Neumann* [6]) Let $X = \{(x, y, z) \in \mathbb{C}^3;\ z^3 = x^2 + y^2.\}$. Then $(X, \mathbf{0})$ has a homotopically nontrivial $\frac{5}{2}$-fast loop. In particular, $(X, 0)$ is not inner Lipschitz conical. Actually, the homotopically nontrivial loop can be presented by: $\gamma(\theta) = (\cos\theta, \sin\theta, 1)$, where $0 \leq \theta \leq 2\pi$. By using the homotopy $H_t(\theta) = (t^{3/2}\cos\theta, t^{3/2}\sin\theta, t); 0 \leq t \leq 1$, we see that γ is a $\frac{5}{2}$-fast loop of X at $\mathbf{0}$.

Separating sets. A subset $Y \subset X$ rectifiable with real dimension $2dim(X, x) - 1$ is said *a local separating set of X at x* if, for some $\epsilon > 0$, we have the following:

a. Y divides $X \cap B(x, \epsilon)$ in at least two connected pieces A and B;
b. Y has null superior density at x;
c. A and B have positive inferior density at x.

Theorem 32 (Birbrair-Fernandes-Neumann [14]) *Existence of a local separating set is an inner bi-Lipschitz invariant of the germs of subanalytic sets. Moreover, if a germ with isolated singularity has a local separating set, then it is not inner Lipschitz conical.*

Next, we use the same previous example to show a germ of complex algebraic surface with isolated singularity which has a local separating set; then we recover this germ is not inner Lipschitz conical.

Example 33 ([6]) Let $X = \{(x, y, z) \in \mathbb{C}^3;\ z^3 = x^2 + y^2.\}$. Let us write $u = x + iy$ and $v = x - iy$; then,

$$Y = \{(u, v, z) \in X\ :|u| = |v|\}$$

divides X in the following two components

$$A = \{(u, v, z) \in X\ :|u| > |v|\} \text{ and } B = \{(u, v, z) \in X\ :|u| < |v|\}.$$

Since A is isometric to B and X has positive density at $\mathbf{0}$. we get A and B has positive density at $\mathbf{0}$. And, since the tangent cone of Y at $\mathbf{0}$ is the z-axis in $\mathbb{C}^3$ (thus it has real dimension 2) and Y has real dimension 3, we get Y has null density at $\mathbf{0}$. It concludes that Y is a local separating set of X at $\mathbf{0}$.

A few years after the creation of fast loops and separation sets,by using a very clever combination of these invariants, in the paper [19], L. Birbrair, W. Neumann and A. Pichon proved that germs of normal complex algebraic surface singularities has a canonical geometric decomposition, so-called *thin-thick decomposition* of the germ which can be codified by a combinatorial object and they arrive in the following result (see [19, 43]).

Theorem 34 (Birbrair-Neumann-Pichon [19]) *Two germs of normal complex algebraic surface singularities are inner Lipschitz homeomorphic if, and only if, their thin-thick decomposition define isomorphic combinatorial codes.*

About the invariants for the problem of the bi-Lipschitz classification of singularities in higher dimensions that have already been found, we highlight the following homology/homotopy theories:

- Metric Homology developed by L. Birbrair and J.-P. Brasselet (see [3, 4]);
- Vanishing homology developed by G. Valette (see [48]);
- Moderately Discontinuous Homology developed by J. de Bobadilla, S. Heinze, M. Pe Pereira and E. Sampaio (see [22]);
- The homotopy theory developed by J. de Bobadilla, S. Heinze and M. Pe-Pereira (see [21]).

We also would like to mention the invariants developed by L. Birbrair, A. Fernandes, V. Grandjeand and D. O'Shea, so-called *choking horns* (see [9]) and the so-called *choking cycles* developed by A. Fernandes and E. Sampaio (see [29]).

We end this section recalling that the problem of the outer bi-Lipschitz classification of complex surface singularities is not solved yet.

6 Global Lipschitz Geometry

Before presenting some global bi-Lipschitz classification of affine subsets $X \subset \mathbb{K}^n$, let us introduce the following crucial notion of bi-Lipschitz homeomorphisms at infinity (see [27]).

Definition 35 Let $X \subset \mathbb{K}^m$ and $Y \subset \mathbb{K}^n$. We say that X is *outer (inner) bi-Lipschitz homeomorphic to Y at infinity* if there are compact subsets $K_1 \subset \mathbb{K}^m$, $K_2 \subset \mathbb{K}^n$ and an outer (inner) bi-Lipschitz homeomorphism $X \cap (\mathbb{K}^m \setminus K_1) \to Y \cap (\mathbb{K}^n \setminus K_2)$.

It is important to mention that, unlike the local case, there are subsets which are diffeomorphic at infinity but not outer bi-Lipschitz homeomorphic at infinity. For instance, we see that

$$\{(x, y) \in \mathbb{C}^2 \ : \ y = x^2\}$$

is diffeomorphic to

$$\{(x, y) \in \mathbb{C}^2 \ : \ y = 0\}$$

via the automorphism $(x, y) \mapsto (x, y - x^2)$; but, those subsets are not outer bi-Lipschitz homeomorphic (at infinity). Next result, obtained in the joint paper with Sampaio [27], shows more evidence of the rigidity of bi-Lipschitz outer equivalence at infinity between affine algebraic sets.

Theorem 36 (Fernandes-Sampaio [27]) *Let $X \subset \mathbb{C}^n$ be an algebraic subset. If X is outer bi-Lipschitz homeomorphic to some d-dimensional vector subspace of $\mathbb{C}^n$ at infinity, then X is a d-dimensional vector subspace of $\mathbb{C}^n$ itself.*

In order to introduce the first global bi-Lipschitz classification result, namely, the outer bi-Lipschitz classification of complex algebraic plane curves by R. Targino (see [47]), let us introduce a collection of discrete invariants associated to these curves.

Definition 37 Let $X \subset \mathbb{C}^2$ be an algebraic plane curve. For each singular point p of X, the Puiseux pairs of each branch of (X, p) and their intersection numbers are said *the Zariski local data of* X *at* p. Also, let us consider an embedding of $\mathbb{C}^2$ into the projective plane $\mathbb{CP}^2$, and let $\overline{X}$ be the respective projective closure of X and let L_∞ be the respective line at infinity in $\mathbb{CP}^2$. For each $q \in \overline{X} \cap L_\infty$, the Puiseux pairs of the branches of $(\overline{X} \cap L_\infty, q)$ and their intersection numbers are said the *the Zariski data of* X *at infinity*.

We are able to state the main results on the paper [47] where is given a global classification of complex algebraic plane curves up to outer bi-Lipschitz homeomorphisms.

Theorem 38 (Targino [47]) *Let* $X \subset \mathbb{C}^2$ *and* $Y \subset \mathbb{C}^2$ *be two complex algebraic curves.*

1. *The curves* X *and* Y *are outer bi-Lipschitz homeomorphic at infinity if, and only if, their Zariski data at infinity are the same;*
2. *The curves* X *and* Y *are outer bi-Lipschitz homeomorphic if, and only if, they have the same Zariski data at infinity and there exists a bijection* $\sigma : Sing(X) \to Sing(Y)$ *such that* $\forall\ p \in Sing(X)$*, the Zariski local data of* X *at* p *is equal to the Zariski local data of* Y *at* $\sigma(p)$.

We end this section, and also these notes, by bringing a recent result, obtained in the joint paper with Sampaio [30], where we use the ideas and local techniques developed by L. Birbrair in the article [2] to give a global classification of properly embedded semialgebraic surfaces in $\mathbb{R}^n$ with isolated singularities up to inner bi-Lipschitz homeomorphisms.

Let $S \subset \mathbb{R}^n$ be a closed semi-algbebraic subset with isolated singularities. Let $Sing(S)$ be the set of the singular points of S. Since S is semi-algebraic and $Sing(S)$ is discrete subset, we get $Sing(S)$ is a finite set. By using an inversion of S in $\mathbb{R}^n$ (let us say the inversion centered at the origin) and the Local Conical Structure Theorem 9, we see that there is a radius $r_0 > 0$ such that for any $r \geq r_0$ the set $S \setminus B_r$, where B_r is the Euclidean ball centered at the origin with radius r, is diffeomorphic to the cylinder $[r, \infty) \times (S \cap \partial B_r)$. The connected components of $S \setminus B_r$ are called *the ends of* S, and any two family of ends corresponding to r and $\tilde{r}$, respectively, are always diffeomorphic each other, for any pairs of radius $r, \tilde{r} > r_0$. Since S is semi-algebraic, it has a finite number of ends, let us say: $E_1, \ldots, E_e$, where $e = e(S)$ is the number of ends of S. For each end E_i of S, it is proved in the paper [30] that the following limit exists and it is a rational number $\alpha_i(S) \leq 1$.

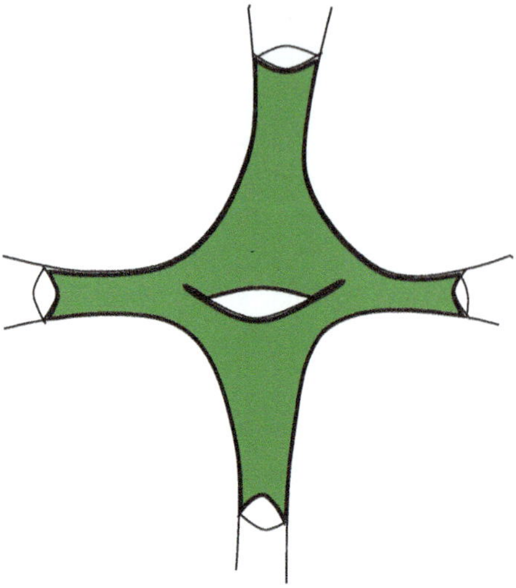

Fig. 2 An orientable and genus 1 semi-algebraic surface with 4 ends and without singularities

$$\lim_{r\to\infty} \frac{\log diam(E_i \cap \partial B_r)}{\log(r)}$$

where $diam(\cdot)$ means the diameter of the set (Fig. 2).

Definition 39 Let $S \subset \mathbb{R}^n$ be a 2-dimensional closed semi-algbebraic subset with isolated singularities. Let us suppose that S has normal singularities in the sense that the link of S at p is connected for any $p \in Sing(S)$. Such a set S we call a *semi-algebraic surface with normal singularities*. We say that the collection of numbers below is the *inner Lipschitz code of* S:

- $g(S)$: genus of S;
- $o(S)$ which is and $+1$ if S is orientable and -1 if not;
- $\beta(S, p_1) \leq \cdots \leq \beta(S, p_\sigma)$ the β-horn exponents given by the Theorem 14 of Birbrair, where $p_1 \ldots, p_\sigma$ are the singular points of S;
- $\alpha(E_1) \leq \cdots \leq \alpha(E_e)$ where $E_1, \ldots, E_e$ are the ends of S.

Theorem 40 (Fernandes-Sampaio [30]) *Let $S \subset \mathbb{R}^n$ and $\tilde{S} \subset \mathbb{R}^m$ be semi-algebraic surfaces with normal singularities. Then, S and $\tilde{S}$ are inner bi-Lipschitz homeomorphic if, and only if, they have the same inner Lipschitz code.*

Remark 41 In the paper [30], the authors defined the inner Lipschitz code of a closed 2-dimensional semi-algebraic subset of $\mathbb{R}^n$ with isolated (not necessarily normal) singularities and they proved Theorem 40 without the assumption the singularities of S and $\tilde{S}$ are normal.

How to classify 2-dimensional real semi-algebraic/subanalytic germs up to outer bi-Lipschitz homeomorphisms?

The first one is the problem of classifying germs of real subanalytic sets of dimension 2 equipped with the outer metric; this problem remains unsolved. I think this problem is among the most difficult problems in Lipschitz geometry that are on the verge of being solved. We have seen that such problem is solved when the germ of surface is Lipschitz normally embedded; it is nothing else than the Theorem 14. So, a significant contribution towards of the case where the germ of subanalytic surface is not Lipschitz normally embedded, for instance, is the case where there is a partition of the germ as two Lipschitz normally embedded pieces (topological triangles) with a technical hypothesis on the relative position of the triangles; this case is address in the preprint paper [16].

How to classify 2-dimensional complex algebraic/analytic germs up to outer bi-Lipschitz homeomorphisms?

The second one is the problem of classifying germs of complex analytic sets of dimension 2 equipped with the outer metric; this problem also remains unsolved. Of course, this problem is solved for germs of complex analytic surfaces which are Lipschitz normally embedded, this is nothing else than Theorem 34. As one can see in the paper [36] where the W. Neumann, H. Pedersen and A. Pichon presented a characterization of 2-dimensional normal complex analytic germs which are Lipschitz normally embedded, the class of such germs which are not Lipschitz normally embedded is quite large. Some invariants towards the outer bi-Lipschitz classification of 2-dimensional normal complex analytic germs are described in the article [43].

References

1. Belotto da Silva, A., Fantini, L., Pichon, A.: On Lipschitz normally embedded complex surface germs. Compos. Math. **158**(3), 623–653 (2022). https://doi.org/10.1112/S0010437X22007357
2. Birbrair, L.: Local bi-Lipschitz classification of 2-dimensional semialgebraic sets. Houston J. Math. **25**(3), 453–472 (1999)
3. Birbrair, L., Brasselet, J.P.: Metric homology. Comm. Pure Appl. Math. **53**(11), 1434–1447 (2000). https://doi.org/10.1002/1097-0312(200011)53:11<1434::aid-cpa5>3.0.co;2-s
4. Birbrair, L., Brasselet, J.P.: Metric homology for isolated conical singularities. Bull. Sci. Math. **126**(2), 87–95 (2002). https://doi.org/10.1016/S0007-4497(01)01085-5

5. Birbrair, L., Costa, J.C.F., Fernandes, A., Ruas, M.A.S.: $\mathcal{K}$-bi-Lipschitz equivalence of real function-germs. Proc. Am. Math. Soc. **135**(4), 1089–1095 (2007). https://doi.org/10.1090/S0002-9939-06-08566-2
6. Birbrair, L., Fernandes, A.: Metric theory of semialgebraic curves. Rev. Mat. Complut. **13**(2), 369–382 (2000). https://doi.org/10.5209/rev_REMA.2000.v13.n2.17080
7. Birbrair, L., Fernandes, A.: Inner metric geometry of complex algebraic surfaces with isolated singularities. Comm. Pure Appl. Math. **61**(11), 1483–1494 (2008). https://doi.org/10.1002/cpa.20244
8. Birbrair, L., Fernandes, A., Grandjean, V., Gabrielov, A.: Lipschitz contact equivalence of function germs in $\mathbb{R}^2$. Ann. Sc. Norm. Super. Pisa Cl. Sci. (5) **17**(1), 81–92 (2017). https://doi.org/10.2422/2036-2145.201503_014
9. Birbrair, L., Fernandes, A., Grandjean, V., O'Shea, D.: Choking horns in Lipschitz geometry of complex algebraic varieties. Appendix by Walter D. Neumann. J. Geom. Anal. **24**(4), 1971–1981 (2014). https://doi.org/10.1007/s12220-013-9405-x
10. Birbrair, L., Fernandes, A., Lê, D.T., Sampaio, J.E.: Lipschitz regular complex algebraic sets are smooth. Proc. Am. Math. Soc. **144**(3), 983–987 (2016). https://doi.org/10.1090/proc/12783
11. Birbrair, L., Fernandes, A., Neumann, W.D.: Bi-Lipschitz geometry of weighted homogeneous surface singularities. Math. Ann. **342**(1), 139–144 (2008). https://doi.org/10.1007/s00208-008-0225-4
12. Birbrair, L., Fernandes, A., Neumann, W.D.: Bi-Lipschitz geometry of complex surface singularities. Geom. Dedicata **139**, 259–267 (2009). https://doi.org/10.1007/s10711-008-9333-2
13. Birbrair, L., Fernandes, A., Neumann, W.D.: On normal embedding of complex algebraic surfaces. In: Manoel, M., Fuster, M.C.R., Wall, C.T.C. (eds.) Real and complex singularities, London Mathematical Society Lecture Note Series, vol. 380, pp. 17–22. Cambridge University Press, Cambridge (2010)
14. Birbrair, L., Fernandes, A., Neumann, W.D.: Separating sets, metric tangent cone and applications for complex algebraic germs. Selecta Math. (N.S.) **16**(3), 377–391 (2010). https://doi.org/10.1007/s00029-010-0024-0
15. Birbrair, L., Gabrielov, A.: Lipschitz geometry of pairs of normally embedded Hölder triangles. Eur. J. Math. **8**(3), 766–791 (2022). https://doi.org/10.1007/s40879-022-00572-2
16. Birbrair, L., Gabrielov, A.: Outer Lipschitz Classification of Normal Pairs of Hölder triangles. arXiv:2311.09461 (2024)
17. Birbrair, L., Mendes, R., Nuño Ballesteros, J.J.: Metrically un-knotted Corank 1 singularities of surfaces in $\mathbb{R}^4$. J. Geom. Anal. **28**(4), 3708–3717 (2018). https://doi.org/10.1007/s12220-017-9973-2
18. Birbrair, L., Mostowski, T.: Normal embeddings of semialgebraic sets. Michigan Math. J. **47**(1), 125–132 (2000). https://doi.org/10.1307/mmj/1030374672
19. Birbrair, L., Neumann, W.D., Pichon, A.: The thick-thin decomposition and the bilipschitz classification of normal surface singularities. Acta Math. **212**(2), 199–256 (2014). https://doi.org/10.1007/s11511-014-0111-8
20. Birbrair, L., Sena, E.: Geometria Lipschitz das singularidades. 33. º Colóquio Brasileiro de Matemática. Instituto Nacional de Matemática Pura e Aplicada (IMPA), Rio de Janeiro (2021)
21. Fernández de Bobadilla, J., Heinze, S., Pe Pereira, M.: Moderately discontinuous homotopy. Int. Math. Res. Not. IMRN **2022**(23), 18346–18400 (2022). https://doi.org/10.1093/imrn/rnab225
22. Fernández de Bobadilla, J., Heinze, S., Pe Pereira, M., Sampaio, J.E.: Moderately discontinuous homology. Comm. Pure Appl. Math. **75**(10), 2123–2200 (2022). https://doi.org/10.1002/cpa.22013
23. Brieskorn, E., Knörrer, H.: Plane algebraic curves. Birkhäuser Verlag, Basel (1986). https://doi.org/10.1007/978-3-0348-5097-1. Translated from the German by John Stillwell
24. Denkowski, M., Tibăr, M.: Testing Lipschitz non-normally embedded complex spaces. Bull. Math. Soc. Sci. Math. Roumanie (N.S.) **62(110)**(1), 93–100 (2019)
25. Fernandes, A.: Topological equivalence of complex curves and bi-Lipschitz homeomorphisms. Michigan Math. J. **51**(3), 593–606 (2003). https://doi.org/10.1307/mmj/1070919562

26. Fernandes, A., Sampaio, J.E.: Tangent cones of Lipschitz normally embedded sets are Lipschitz normally embedded. Appendix by Anne Pichon and Walter D. Neumann. Int. Math. Res. Not. IMRN **2019**(15), 4880–4897 (2019). https://doi.org/10.1093/imrn/rnx290
27. Fernandes, A., Sampaio, J.E.: On Lipschitz rigidity of complex analytic sets. J. Geom. Anal. **30**(1), 706–718 (2020). https://doi.org/10.1007/s12220-019-00162-x
28. Fernandes, A., Sampaio, J.E.: Bi-Lipschitz invariance of the multiplicity. In: Cisneros-Molina, J.L., Düng Tráng, L., Seade, J. (eds.) Handbook of geometry and topology of singularities IV, pp. 463–496. Springer, Cham (2023). https://doi.org/10.1007/978-3-031-31925-9_9
29. Fernandes, A., Sampaio, J.E.: On characterization of smoothness of complex analytic sets. Indiana Univ. Math. J. **72**(6), 2547–2565 (2023)
30. Fernandes, A., Sampaio, J.E.: Global bi-Lipschitz classification of semi-algebraic surfaces. Ann. Sc. Norm. Super. Pisa Cl. Sci. (5) **25**(3), 1505–1526 (2024). https://doi.org/10.2422/2036-2145.202206_005
31. Fernandes, A., Sampaio, J.E., Silva, J.P.: Hölder equivalence of complex analytic curve singularities. Bull. Lond. Math. Soc. **50**(5), 874–886 (2018). https://doi.org/10.1112/blms.12192
32. Kerner, D., Pedersen, H.M., Ruas, M.A.S.: Lipschitz normal embeddings in the space of matrices. Math. Z. **290**(1–2), 485–507 (2018). https://doi.org/10.1007/s00209-017-2027-4
33. Mendes, R., Sampaio, J.E.: On the link of Lipschitz normally embedded sets. Int. Math. Res. Not. IMRN **2024**(9), 7488–7501 (2024). https://doi.org/10.1093/imrn/rnad215
34. Misev, F., Pichon, A.: Lipschitz normal embedding among superisolated singularities. Int. Math. Res. Not. IMRN **2021**(17), 13546–13569 (2021). https://doi.org/10.1093/imrn/rnz221
35. Mostowski, T.: Lipschitz equisingularity. Dissertationes Math. (Rozprawy Mat.) **243**, 46 (1985)
36. Neumann, W.D., Pedersen, H.M., Pichon, A.: A characterization of Lipschitz normally embedded surface singularities. J. Lond. Math. Soc. (2) **101**(2), 612–640 (2020). https://doi.org/10.1112/jlms.12279
37. Neumann, W.D., Pedersen, H.M., Pichon, A.: Minimal surface singularities are Lipschitz normally embedded. J. Lond. Math. Soc. (2) **101**(2), 641–658 (2020). https://doi.org/10.1112/jlms.12280
38. Neumann, W.D., Pichon, A.: Lipschitz geometry of complex curves. J. Singul. **10**, 225–234 (2014). https://doi.org/10.5427/jsing.2014.10o
39. Neumann, W.D., Pichon, A. (eds.): Introduction to Lipschitz geometry of singularities, Lecture Notes in Mathematics, vol. 2280. Springer, Cham (2020). https://doi.org/10.1007/978-3-030-61807-0. Lecture notes of the International School on Singularity Theory and Lipschitz Geometry, Cuernavaca (2018)
40. Parusiński, A.: Lipschitz properties of semi-analytic sets. Ann. Inst. Fourier (Grenoble) **38**(4), 189–213 (1988). https://doi.org/10.5802/aif.1154
41. Pham, F., Teissier, B.: Fractions Lipschitziennes d'une algèbre analytique complexe et saturation de Zariski, par Frédéric Pham et Bernard Teissier (1969). https://hal.science/hal-00384928. Centre de Mathématiques de l'Ecole Polytechnique (Paris). 42 p. Ce travail est la base de l'exposé de Frédéric Pham au Congrès International des Mathématiciens, Nice (1970)
42. Pham, F., Teissier, B.: Lipschitz fractions of a complex analytic algebra and Zariski saturation. In: Neumann, W., Pichon, A. (eds.) Introduction to Lipschitz geometry of singularities, Lecture Notes in Mathematics, vol. 2280, pp. 309–337. Springer, Cham (2020). https://doi.org/10.1007/978-3-030-61807-0_10
43. Pichon, A.: An introduction to Lipschitz geometry of complex singularities. In: Neumann, W.D., Pichon, A. (eds.) Introduction to Lipschitz geometry of singularities, Lecture Notes in Mathematics, vol. 2280, pp. 167–216. Springer, Cham (2020). https://doi.org/10.1007/978-3-030-61807-0_7
44. Ruas, M.A.S.: Basics on Lipschitz geometry. In: Neumann, W.D., Pichon, A. (eds.) Introduction to Lipschitz geometry of singularities, Lecture Notes in Mathematics, vol. 2280, pp. 111–155. Springer, Cham (2020). https://doi.org/10.1007/978-3-030-61807-0_5
45. Sampaio, J.E.: Bi-Lipschitz homeomorphic subanalytic sets have bi-Lipschitz homeomorphic tangent cones. Selecta Math. (N.S.) **22**(2), 553–559 (2016). https://doi.org/10.1007/s00029-015-0195-9

46. Sampaio, J.E.: Some classes of homeomorphisms that preserve multiplicity and tangent cones. In: Castro-Jiménez, F.J., Massey, D.B., Teissier, B., Tosun, M. (eds.) A panorama of singularities, Contemporary Mathematics, vol. 742, pp. 189–200. American Mathematical Society [Providence], RI (2020). https://doi.org/10.1090/conm/742/14945
47. Targino, R.: Outer Lipschitz geometry of complex algebraic plane curves. Int. Math. Res. Not. IMRN **2023**(15), 13255–13289 (2023). https://doi.org/10.1093/imrn/rnac202
48. Valette, G.: Vanishing homology. Selecta Math. (N.S.) **16**(2), 267–296 (2010). https://doi.org/10.1007/s00029-010-0020-4

The Role of Resampling Methods and Generalized Means in Extreme Value Theory

M. Ivette Gomes

Abstract Resampling methodologies have recently revealed to be very fruitful in the field of *statistics of extremes*. First, we mention the importance of the *Generalized Jackknife* to reduce bias. We next refer to the relevance of the *Bootstrap* in the estimation of a crucial tuning parameter in the area, related to the number k of upper-order statistics involved in the estimation of tails. Moreover, most of the estimators of parameters of rare and large events, among which we distinguish the *extreme value index* (EVI), are averages of statistics, based on k. Only for heavy tails, quite common in many areas of application, classes of reliable estimators of parameters of rare events based on adequate *generalized means* (GMs) are introduced and discussed. Together, these two resampling procedures and GMs enable the obtention of reliable semi-parametric estimates of any parameter of extreme events, like the primary parameter of extreme events, the EVI.

1 Introduction

Apart from the estimation of the *extreme value index* (EVI), the reliable estimation of other important parameters of rare events will be among the topics to be addressed, giving emphasis to the use of resampling methodologies, like the *Generalized Jackknife* (GJ), to reduce bias, and the Bootstrap, to adequately choose the high threshold. The role of *generalized means* (GMs) and of the *peaks over a random threshold* (PORT) methodology, in a reliable estimation of the EVI will also be discussed. Indeed, the EVI is of general interest in EV *theory* (EVT), regardless of the framework one considers, being the key for different methodologies, like the block maxima (Gumbel [72]), the multivariate and multidimensional EV models (Gomes [31, 32]), the POT approach (Smith [95]), as well as for novel EV models that are based on all data—and not just on the extremes [20, 82]. In fact, even some

M. I. Gomes (✉)
CEAUL and DEIO, FCUL, Universidade de Lisboa, Academia das Ciências de Lisboa, IIC Bento da Rocha Cabral, Lisboa, Portugal
e-mail: ivette.gomes@ciencias.ulisboa.pt

M. Abreu et al. (eds.), *Brazil-Portugal Mathematics*, Coimbra Mathematical Texts 4,
https://doi.org/10.1007/978-3-031-98455-6_6

parameters for multivariate extremes, like the tail dependence coefficient (see, [21, 38, 80], among others), are based on the EVI of an univariate *random variable* (RV). Among review papers in the area of statistics of extremes, and apart from [21], we mention [3, 24, 43, 50, 70, 79, 83, 93]. After a brief motivation, in Sect. 2, for the need of EVT, in Sect. 3, and for a heavy right-tail, i.e. a positive EVI, the most common EVI-estimators, the Hill (H) EVI-estimators (Hill [78]), are introduced. These estimators are dependent on a tuning parameter k, related to the number of *order statistics* (OSs) used in the estimation, and, unless the underlying model is a Pareto model, they exhibit a high variance for small k and a high bias for large k. Resampling methodologies have revealed to be very fruitful in the finding of solutions for the aforementioned problems, and will be briefly discussed in Sect. 4. We there mention the importance of the GJ and of the *Bootstrap* in the obtention of a reliable semi-parametric estimate of any parameter of extreme events, like the EVI. To illustrate the GJ methodology, we shall consider the simplest class of *minimum-variance reduced-bias* (MVRB) EVI-estimators of a positive EVI, introduced in [18]. Further note that, like most estimators in the general area of Statistics, the H EVI-estimators are the average of adequate statistics, based on the k upper observations. Trying to improve the performance of the classical H EVI-estimators, recent classes of reliable EVI-estimators based on adequate GMs are introduced and discussed in Sect. 5, individually and together with the use of the PORT methodology. Finally, in Sect. 6, a few overall comments are provided, together with some open research topics.

2 Motivation for the Need of EVT

EVT helps us to control potentially disastrous events, of high relevance to society and with a high social impact, being indeed a mathematical decoder of the random world of extremes. Domains of application of EVT are quite diverse. See, among other books, [6, 22, 23, 48, 81, 92], where different case-studies are presented.

Statistics of univariate extremes, as well as multivariate and spatial extremes, have recently faced a huge development, partially due to the fact that extreme events can have catastrophic consequences for human activities, through their impact not only on natural but also on constructed environments. Parametric modeling, based on probabilistic asymptotic results in EVT, was the usual statistical EVT framework until the late seventies. In the eighies there has been a shift from the area of parametric *statistics of extremes*, towards semi-parametric or even non-parametric approaches. But parametric modeling is becoming again quite popular, particularly in spatial applications of EVT.

To motivate the interest for this area, we merely refer the North Sea Historical Floods (see Fig. 1). As a consequence of these floods, the Dutch government formed the so-called '*Delta Committee*'. It was decided that dams should be built in such a way that *the probability of a flood in a certain year should be* 1 *out of* 10 000. But data have been observed during a much shorter period. It is thus necessary to extrapolate much beyond observed data. And, on the basis of the so-called return

Fig. 1 North-sea floods: According to Encyclopedia Britannica [97], at the early morning of February 1, 1953, the level of the water exceeded 5.6 m above the sea level, the maritime defenses were destroyed, provoking the flooding of areas in Holland, England, Belgium, Denmark and France, with the death of around 2500 people

period of an extreme event, EVT can provide reliable answers on the height of such dam (not always taken into account due to high costs). Such a return period is merely the mean time interval between consecutive occurrences of a certain extreme event, like the floods in the North Sea. For other examples, see the aforementioned books.

Indeed, and just as mentioned in most diverse articles, EVT is needed because the world is not always *normal*, a model with quadratic exponential right and left tails. And risky events are in the tails of the underlying distribution, with most of the data in the center, being observations in the tail quite scarce. Estimates much beyond the observed maximum/minimum are often required, and we thus need to consider models for the tail, usually based on asymptotic results. We can thus put the question whether it is sensible to use those models in all real situations dealing with extreme events. And, being next citing George Box (see [9], p. 424), we cannot forget that '*...all models are wrong but some models are useful.*'

Anyway, the answer to the question whether there is a hidden pattern underlying this type of extreme events is positive, and the most relevant univariate asymptotic models are the *general extreme value* (GEV) and the *generalized Pareto* (GP) models, which are defined in Sect. 3. For other relevant asymptotic models, and different approaches to statistics univariate of extremes, see the overviews in [3, 24, 50, 93].

3 The Class of Hill's EVI-Estimators and Bias Reduction

After a possible adequate transformation, let us consider that $\underline{X}_n = (X_1, \ldots, X_n)$ can be regarded as a sample of size n of *independent, identically distributed* (IID) or more generally stationary weakly dependent RVs from a *cumulative distribution function* (CDF) F. Let us denote by $X_{1:n} \leq \cdots \leq X_{n:n}$ the associated ascending OSs. Further assume that there exist sequences of real constants $\{a_n > 0\}$ and $\{b_n \in \mathbb{R}\}$ such that the maximum, linearly normalized, i.e. $(X_{n:n} - b_n)/a_n$, converges in distribution to a non-degenerate RV. Then, Fisher and Tippett [28] (see also, Gnedenko [30]) have shown that the limiting CDF is necessarily of the type of the GEV CDF, given by

$$G_\xi(x) \equiv \mathrm{GEV}_\xi(x) := \begin{cases} \mathrm{e}^{-(1+\xi x)^{-1/\xi}}, \; 1+\xi x > 0, \text{ if } \xi \neq 0, \\ \mathrm{e}^{-\mathrm{e}^{-x}}, \; x \in \mathbb{R}, \qquad\qquad \text{if } \xi = 0. \end{cases} \tag{1}$$

The CDF F is then said to belong to the *max-domain of attraction* of G_ξ, and we write

$$F \in \mathcal{D}_{\mathcal{M}}\left(G_\xi\right) \quad \text{or} \quad F \in \mathcal{D}_{\mathcal{M}}\left(\mathrm{GEV}_\xi\right).$$

The parameter ξ is the EVI, the primary parameter of extreme events, which measures the heaviness of the *right-tail function*

$$\overline{F}(x) := 1 - F(x),$$

and the heavier the right-tail, the larger ξ is (see Fig. 2, for an illustration of the right-tail heaviness).

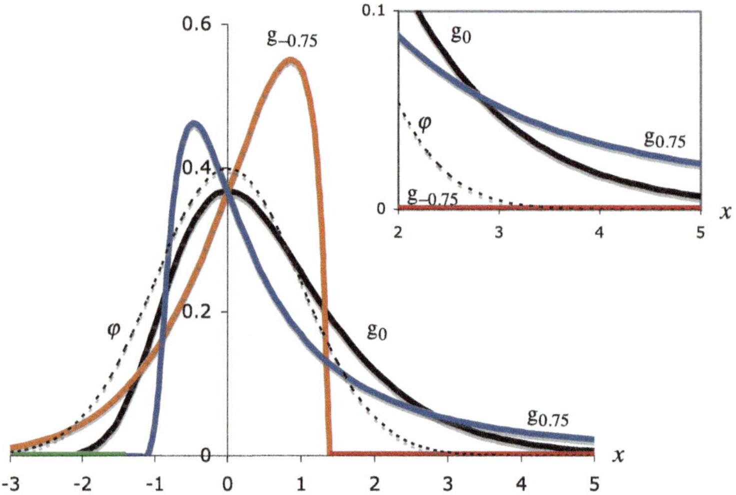

Fig. 2 Probability density function (PDF), $g_\xi = G'_\xi$, with G_ξ given in (1), for $\xi = -0.75$, $\xi = 0$ and $\xi = 0.75$, together with the normal PDF, $\varphi(x) = \exp(-x^2/2)/\sqrt{2\pi}$, $x \in \mathbb{R}$

Rather than the maxima, it is also common to consider all values larger than a given threshold, and the differences between these values and such a threshold, the so-called *exceedances over the threshold*. These exceedances are typically assumed to have a $\mathrm{GP}_\xi(x/\sigma) =: \mathrm{GP}_{\xi,\sigma}(x)$ CDF, $\sigma > 0$, with GP_ξ the GP CDF, defined by

$$\mathrm{GP}_\xi(x) \equiv \mathrm{GP}_{\xi,1}(x) := 1 + \ln \mathrm{GEV}_\xi(x), \tag{2}$$

with $\mathrm{GEV}_\xi(x)$ given in (1). Indeed, and according to [2, 91], the key result in EVT that explains the importance of the GP CDF is the fact that

$$F \in \mathcal{D}_M(\mathrm{GEV}_\xi) \iff \lim_{u \to x^F} \sup_{x \in [0, x^F - u)} \left| F_u(x) - \mathrm{GP}_\xi(x/\sigma(u)) \right| = 0$$

for some positive function $\sigma(u)$, where

$$F_u(x) := \mathbb{P}\big(X - u \le x \big| X > u\big),$$

for $0 \le x \le x^F - u$, is the conditional CDF of excesses over a high threshold u. Thus, the GP distribution is the natural model for the unknown excess distribution above sufficiently high thresholds. This approach, based on the GP approximation in (2), is called the POT approach. Apart from the CDFs in (1) and (2), several other models can appear in an EVT framework (see [10], a recent article where some of those models are discussed).

Despite of the fact that similar results are valid for a general tail, we shall consider *heavy-tailed* models, i.e. *Pareto-type* underlying CDFs, with a positive EVI, working thus in

$$\mathcal{D}_{\mathcal{M}}^+ := \mathcal{D}_{\mathcal{M}}\left(\mathrm{GEV}_{\xi>0}\right),$$

with GEV_ξ defined in (1). These heavy-tailed models are quite common in many areas of application, like biostatistics, computer science, dynamical systems, environment, finance, insurance, statistical quality control, structural engineering and telecommunications, among others.

Note that, with $\mathcal{R}_\alpha$ denoting the class of regularly varying functions with an index of regular variation α, i.e. positive measurable functions $g(\cdot)$ such that $\forall x > 0$, $g(tx)/g(t) \to x^\alpha$, as $t \to \infty$ (see [8], for detais on regular variation),

$$F \in \mathcal{D}_{\mathcal{M}}^+ \iff \overline{F} = 1 - F \in \mathcal{R}_{-1/\xi}. \tag{3}$$

Moreover, and sligthtly more restrictively than (3), it is often assumed that, for all $x > 0$, and as $t \to +\infty$,

$$\overline{F}(tx)/\overline{F}(t) - x^{-1/\xi} = O\big(\beta t^{\rho/\xi}\big), \tag{4}$$

where $\beta \ne 0$ and $\rho < 0$ are respectively generalized scale and shape second-order parameters, a class of models introduced in [74].

For heavy-tailed models, and just as mentioned before, the classical EVI-estimators are the Hill (H) estimators, introduced and studied in [78]. Such EVI-estimators are the average of the log-excesses, i.e.

$$\mathrm{H}(k) = \mathrm{H}(k; \underline{X}_n) := \frac{1}{k}\sum_{i=1}^{k} V_{ik}, \quad V_{ik} := \ln \frac{X_{n-i+1:n}}{X_{n-k:n}}, \quad 1 \le i \le k < n. \tag{5}$$

Consistency is achieved in the whole $\mathcal{D}_{\mathcal{M}}^{+}$, provided that $k = k_n$ is *intermediate*, i.e.

$$k = k_n \to \infty \quad \text{and} \quad k_n/n \to 0, \ \text{as} \ n \to \infty. \tag{6}$$

To obtain asymptotic normality, we need to strengthen the aforementioned condition (3) into a second-order one, possibly of the form (4). Then,

$$\sqrt{k}\,(\mathrm{H}(k) - \xi) \stackrel{d}{=} \mathrm{Normal}_{0,\xi^2} + O_p\big(\sqrt{k}(n/k)^{\rho}\big).$$

Assuming that $\sqrt{k}(n/k)^{\rho} \to \lambda$ finite, asymptotic normality is valid, but with a non-null asymptotic bias whenever $\lambda \neq 0$. Indeed, contrarily to the variance, ξ^2/k, which decreases to zero with k, the bias term, of the order of $\sqrt{k}(n/k)^{\rho}$, $\rho \le 0$, increases drastically as k increases and there is an important trade-off between bias and variance, linked to the choice of the threshold or the number of OSs used, i.e. of the tuning parameter k. Resampling methodologies have recently revealed to be very fruitful in the finding of solutions for the aforementioned problems, and will be briefly discussed in Sect. 4.

Before the development of Sect. 4, we still refer that a direct reduction of the bias of Hill's EVI-estimators in Hall-Welsh class of models, in (4), has led the authors in [18] to the MVRB EVI-estimators,

$$\overline{\mathrm{H}}(k) = \overline{\mathrm{H}}_{\hat\beta,\hat\rho}(k) := \mathrm{H}(k)\big(1 - \hat\beta(n/k)^{\hat\rho}/(1-\hat\rho)\big) \tag{7}$$

with $\mathrm{H}(k)$ the Hill EVI-estimators, in (5), and where $\hat\beta$ and $\hat\rho$ need to be adequate consistent estimators of scale and shape second-order parameters β and ρ. Under the same conditions as before, i.e. (4) and (6), $\sqrt{k}\big(\overline{\mathrm{H}}(k) - \xi\big)$ is asymptotically normal with variance also equal to ξ^2 but with a null mean value. Indeed, from the results in [18], we know that it is possible to adequately estimate the second-order parameters β and ρ, so that we get

$$\sqrt{k}\big(\overline{\mathrm{H}}(k) - \xi\big) \stackrel{d}{=} \mathcal{N}(0, \xi^2) + o_p\big(\sqrt{k}A(n/k)\big).$$

Consequently, $\overline{\mathrm{H}}(k)$ outperforms $\mathrm{H}(k)$ for all k. Just as advised before in [16], we suggest the use of the ρ-estimators in [29] and the β-estimators in [58]. But more recent classes of ρ-estimators (see [76], among others) and β-estimators (see [77], also among others), where the PORT methodology is considered, are potential candidates for the (β, ρ)-estimation. Overviews on reduced-bias estimation can be found in Sect. 6.6 of [92] (see also, [3, 50]).

4 Resampling Methodologies

The use of resampling methodologies (Efron [26]) has revealed to be promising in the estimation of the tuning parameter k, and in the reduction of bias of any estimator of a parameter of extreme events. If we ask how to choose the tuning parameter k in the estimation of a parameter of extreme events, η, through $T(k)$, we usually consider the estimation of

$$k_0^T := \arg\min_k \text{MSE}(T(k)), \tag{8}$$

where MSE stands for *mean squared error*. To obtain estimates of k_0^T, in (8), one can then use a *double-Bootstrap* method applied to an adequate *auxiliary statistic* like

$$A(k) := T(k) - T(\lfloor k/2 \rfloor), \tag{9}$$

where $\lfloor x \rfloor$ stands as usual to the integer part of x. For adequate values of k, the auxiliary statistic $A(k)$ tends to the known value *zero* and has an asymptotic behaviour similar to the one of $T(k)$ (see [37, 65], among others). And algorithms for the Bootstrap estimation of k_0^T, not sketched here, can be easily developed. See, [14, 35, 36, 42, 47, 51, 84], also among others.

At such optimal levels, in the sense of minimal MSE, we can have a non-null asymptotic bias. If we still want to remove such a bias, we can then make use of the GJ methodology. The main objectives of the *Jackknife methodology* are bias and variance estimation of a certain statistic, only through manipulation of observed data $\underline{x}$, and also the building of estimators with bias and MSE smaller than those of an initial set of estimators.

The *Jackknife* or the GJ are resampling methodologies, which usually give a positive answer to the question whether the combination of information may improve the quality of estimators of a certain parameter or functional. Indeed, it is enough to consider an adequate pair of estimators of the parameter of extreme events under study, possibly also $T(k)$ and $T(\lfloor k/2 \rfloor)$, just as in (9), and to built a *reduced-bias affine combination* of them. In [33, 34, 59, 60, 64, 67], also among others, applications of this technique to the H EVI-estimators, in (5), can be found.

4.1 The Jackknife Methodology and Bias Reduction

Most of the pioneering RB EVI-estimators (see [33, 34, 59], already mentioned, and [4, 27, 90]) are, in a certain sense, GJ EVI-estimators, i.e., affine combinations of well-known estimators of ξ.

The GJ statistic was introduced by Gray and Shucany [71]: Let $T_n^{(1)}$ and $T_n^{(2)}$ be two biased estimators of an unknown parameter or functional η, with similar bias properties, i.e.

$$\text{Bias}(T_n^{(i)}) = \eta + \phi(\eta) d_i(n), \quad i = 1, 2.$$

Then, if $q = q_n = d_1(n)/d_2(n) \neq 1$, the affine combination

$$T_n^G := \big(T_n^{(1)} - qT_n^{(2)}\big)/(1-q)$$

is an unbiased estimator of η.

4.2 GJ Corrected-Bias EVI-Estimators

The GJ MVRB EVI-estimators were already studied in [45, 62]. Given the EVI-estimator $\overline{\mathrm{H}}$, in (7) (or any other estimator of a parameter of extreme events), the most natural GJ RV is the one associated with the random pair $\big(\overline{\mathrm{H}}(k), \overline{\mathrm{H}}(\lfloor\theta k\rfloor)\big)$, $0 < \theta < 1$, i.e.

$$\overline{\mathrm{H}}^{\mathrm{GJ}(q,\theta)}(k) := \frac{\overline{\mathrm{H}}(k) - q\,\overline{\mathrm{H}}(\lfloor\theta k\rfloor)}{1-q}, \quad 0 < \theta < 1, \tag{10}$$

with

$$q = q_n = \frac{Bias_\infty[\overline{\mathrm{H}}(k)]}{Bias_\infty[\overline{\mathrm{H}}(\lfloor\theta k\rfloor)]} \underset{n/k\to\infty}{\longrightarrow} \theta^{2\rho}.$$

It is thus sensible to consider, in (10), $q = \theta^{2\rho}$, $\theta = 1/2$, and, with $\hat\rho$ a consistent estimator of ρ, the GJ EVI-estimator,

$$\overline{H}^{\mathrm{GJ}}(k) := \frac{2^{2\hat\rho}\,\overline{H}(k) - \overline{H}(\lfloor k/2\rfloor)}{2^{2\hat\rho} - 1}. \tag{11}$$

Then, and provided that $\widehat{\rho} - \rho = o_p(1)$,

$$\sqrt{k}\big(\overline{H}^{\mathrm{GJ}}(k) - \xi\big) \overset{d}{=} \mathrm{Normal}_{0,\sigma^2_{\mathrm{GJ}}} + o_p\big(\sqrt{k}(n/k)^{2\rho}\big),$$

with

$$\sigma^2_{\mathrm{GJ}} = \xi^2(1 + 1/(2^{-2\rho} - 1)^2 > \sigma^2_{\mathrm{H}} = \sigma^2_{\overline{\mathrm{H}}} = \xi^2.$$

We have thus again a trade-off between variance and bias. The bias decreases and the variance increases. But we are able to reach a better performance at optimal levels.

4.3 Asymptotic Bias and Efficiency of an Affine Combination of Corrected-Bias EVI-Estimators

Just as mentioned before, the most obvious affine combination associated with the MVRB EVI-estimators $\overline{\mathrm{H}}(k)$, is

$$\overline{\mathrm{H}}^{\mathrm{GJ}(a)}(k) \equiv \overline{\mathrm{H}}^{\mathrm{GJ}(a)}(k;\theta) =: a\overline{\mathrm{H}}([\theta k]) + (1-a)\overline{\mathrm{H}}(k),\ \theta = 1/2. \qquad (12)$$

We have thus a class of EVI-estimators parameterized in the *tuning* parameter a, to be chosen in the most adequate way. Note that $\overline{\mathrm{H}}^{\mathrm{GJ}(q,\theta)}(k)$, in (10), has the functional form of $\overline{\mathrm{H}}^{\mathrm{GJ}(a)}(k;\theta)$, in (12), with $a = q/(q-1) = \theta^{2\rho}/(\theta^{2\rho}-1) > 1$.

We shall next analyze the behaviour of the asymptotic bias and efficiency of the affine combination, $\overline{\mathrm{H}}^{\mathrm{GJ}(a)}(k)$, in (12), studying first their asymptotic properties for a possibly non-optimal choice of $a > 1$. For a fixed level k, the reduction in the asymptotic bias of $\overline{\mathrm{H}}^{\mathrm{GJ}(a)}(k)$ comparatively with $\overline{\mathrm{H}}(k)$ is measured by the indicator:

$$\mathrm{ABR}_a := \lim_{n\to\infty}\left(\left|\frac{\mathrm{Bias}_\infty\left(\overline{\mathrm{H}}(k)\right)}{\mathrm{Bias}_\infty\left(\overline{\mathrm{H}}^{\mathrm{GJ}(a)}(k)\right)}\right|\right) = \frac{1}{\left|1-a(1-2^{2\rho})\right|},$$

pictured in Fig. 3.

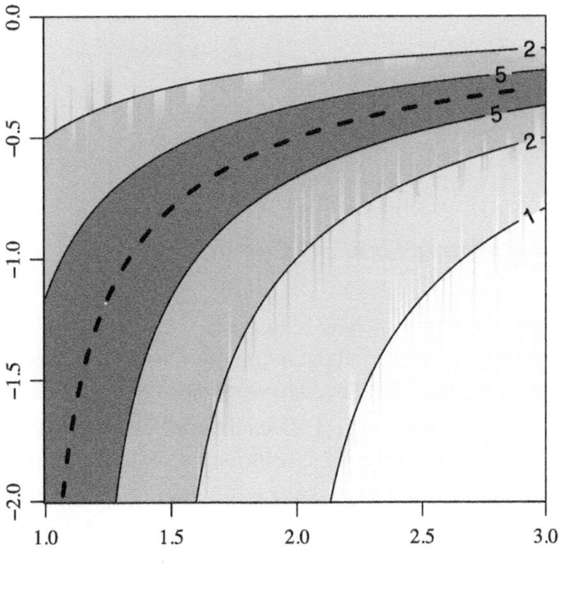

Fig. 3 Asymptotic bias reduction indicator as a function of (a, ρ), $a \geq 1, \rho \leq 0$

Let us next define the asymptotic efficiency of $\overline{\mathrm{H}}^{\mathrm{GJ}(a)}(k)$ relatively to $\overline{\mathrm{H}}(k)$ as the quotient between the two asymptotic MSEs, computed at optimal levels. Provided that $a \neq 1/(1-2^{2\rho})$, we have the indicator

$$\mathrm{AREF}_a := \frac{\mathrm{MSE}_\infty\big(\overline{\mathrm{H}}(k_0^{\overline{\mathrm{H}}})\big)}{\mathrm{MSE}_\infty\big(\overline{\mathrm{H}}^{\mathrm{GJ}(a)}(k_0^{\overline{\mathrm{H}}^{\mathrm{GJ}(a)}})\big)} = \left(\frac{(a^2+1)^{2\rho}}{1-a(1-2^{2\rho})}\right)^{\frac{2}{1-4\rho}}.$$

We next show, in Fig. 4, and again in the (a,ρ)-plane, the values of this asymptotic efficiency, which becomes infinity for $a = 1/(1-2^{2\rho})$, the GJ RV.

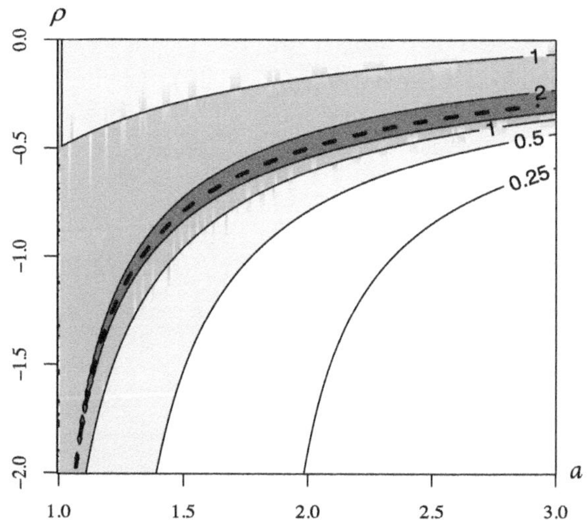

Fig. 4 Asymptotic relative efficiency indicator as a function of (a,ρ), $a \geq 1, \rho \leq 0$

4.4 *Some General Comments*

The reduction in bias holds in a wide region of the (a,ρ)-plane, making almost irrelevant a choice of the *tuning parameter* a. However, it is clear that for a reduction in MSE, we indeed need to work close to the line $a = 1/(1-2^{2\rho})$. This justifies the introduction of the GJ EVI-estimators, $\overline{\mathrm{H}}^{\mathrm{GJ}}(k)$, in (11). There is then a high reduction in the MSE of the GJ $\overline{\mathrm{H}}$-estimators, at their optimal level, comparatively with the original $\overline{\mathrm{H}}$-estimators, also at optimal level, always in the sense of minimal MSE. The sample paths of these corrected-bias estimators are usually quite stable. The choice of the optimal level is thus of a smaller importance. But, even so, we can use the *Bootstrap methodology* for the choice of such an optimal level.

4.5 The GJ EVI-Estimation Applied to Insurance Data

We consider an illustration of the performance of the EVI-estimates under study, through the analysis of automobile claim amounts exceeding 1,200,000 Euro over the period 1988–2001, gathered from several European insurance companies co-operating with the same re-insurer (Secura Belgian Re). This data set was already studied in [6, 45], already mentioned, and in [5, 61, 98], among others, as an example to excess-of-loss reinsurance rating and heavy-tailed distributions in car insurance.

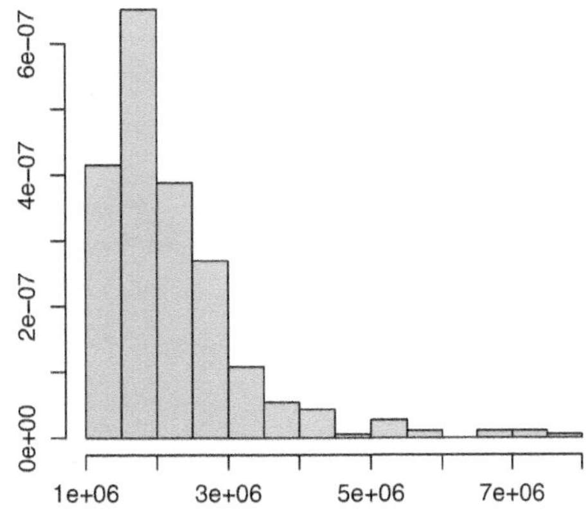

Fig. 5 Histogram associated with the Secura data

A short preliminary graphical analysis of the data, x_i, $1 \leq i \leq n$, $n = 371$, was performed. And it is clear from the histogram in Fig. 5, that data have been left-censored and that the right tail of the underlying model is quite heavy.

See also the extensive prelimary analysis of this data set in [6], where an exponential QQ-plot enables the immediate detection of a point of inflection with different slopes to the left and to the right. As mentioned in the aforementioned book, this becomes even more apparent in the mean excess plot, where beyond 2,500,000 the almost horizontal behaviour changes into a positive slope. Moreover, the Pareto quantile plot is approximately linear in the largest observations, indicating a good fit of a Pareto-type model to the right-tail of the claim size distribution, i.e. the existence of a positive EVI.

Regarding the EVI-estimation, note that whereas the Hill estimator is unbiased for the estimation of ξ when the underlying model is a strict Pareto model, it always exhibits a relevant bias when we have only Pareto-like tails, as surely happens here. The corrected-bias estimators, which are "asymptotically unbiased", have a smaller bias, exhibit more stable sample paths as functions of k, and enable us to take a decision upon the estimate of ξ to be used, even with the help of any heuristic stability criterion, like the "largest run" suggested in [44].

For the Hill estimator, we have simple techniques to estimate the optimal sample fraction (Hall [73]). The Bootstrap algorithm described in [68] (see also, [69]), and not detailed here, helps us to provide an adaptive choice not only for the H but also for any RB EVI-estimator, and was also used for the GJ EVI-estimation. We have got $\hat{k}_{0|\mathrm{H}} = 56$, $\hat{k}_{0|\overline{\mathrm{H}}} = 158$, $\hat{k}_{0|\overline{\mathrm{H}}^{\mathrm{GJ}}} = 261$, and the EVI-Bootstrap estimates

$$\mathrm{H}^{**} = 0.286, \quad \overline{\mathrm{H}}^{**} = 0.240 \quad \text{and} \quad \overline{\mathrm{H}}^{\mathrm{GJ}^{**}} = 0.236,$$

the values pictured in Fig. 6.

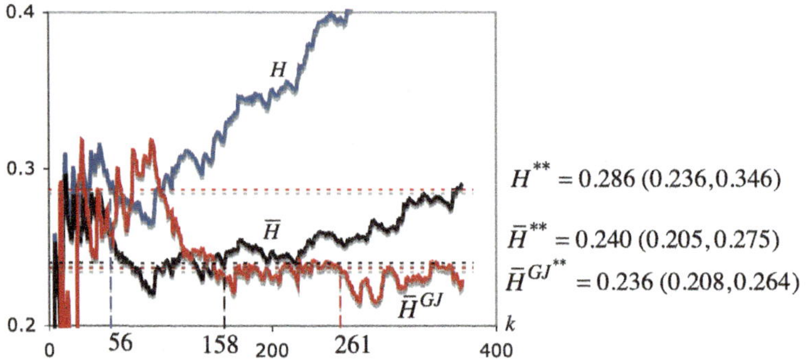

Fig. 6 Estimates of the EVI, ξ, for the Secura Belgian Re data

Remark 1 Note that Bootstrap *confidence intervals* (CIs), also provide in Fig. 6, for a level 95%, as well as asymptotic CIs, are easily associated with the estimates presented, the smallest size (with a high coverage probability) being related to $\overline{\mathrm{H}}^{\mathrm{GJ}^{**}}$.

And GJ semi-parametric Weisssman-type estimatss of high quantiles (Weissman [99]), based on RB-GJ EVI-estimators, exhibit a similar reliable behaviour.

4.6 *Some Conclusions*

The most attractive features of the GJ EVI-estimators are their stable sample paths, close to the target value, for a wide region of large k-values. The optimal choice of k is always larger than the one achieved for the associated RB EVI-estimator, leading thus to a smaller CI, and consequently to a more reliable estimation. We further mention the "bath-tube" MSE patterns of GJ EVI-estimators.

5 Classes of GMs' EVI-Estimators

Being the average of the log-excesses V_{ik} in (5), the H EVI-estimators can be written as

$$\mathrm{H}(k) = \sum_{i=1}^{k} \ln\left(\frac{X_{n-i+1:n}}{X_{n-k:n}}\right)^{1/k} = \ln\left(\prod_{i=1}^{k} \frac{X_{n-i+1:n}}{X_{n-k:n}}\right)^{1/k} =: \ln\left(\prod_{i=1}^{k} U_{ik}\right)^{1/k}. \quad (13)$$

The H EVI-estimators are thus the logarithm of the *geometric mean* (or *mean-of-order*-0) of U_{ik}, $1 \le i \le k$, defined in (13).

5.1 *Hölders's Mean-of-Order-p (H_p) EVI-Estimators*

The authors in [11], and almost simultaneously the ones in [7, 85], considered as basic statistics, the *power mean-of-order-p* (MO_p) of U_{ik}, $1 \le i \le k$, in (13), for $p \ge 0$ (see also [12]). More generally, (see [16, 40]), those same statistics were considered for any $p \in \mathbb{R}$, i.e.

$$\mathrm{MO}_p(k) = \begin{cases} \left(\frac{1}{k}\sum_{i=1}^{k} U_{ik}^p\right)^{1/p}, & \text{if } p \neq 0, \\ \left(\prod_{i=1}^{k} U_{ik}\right)^{1/k}, & \text{if } p = 0, \end{cases}$$

and the associated class of EVI-estimators,

$$\mathrm{H}_p(k) = \mathrm{H}_p(k; \underline{X}_n) := \begin{cases} \left(1 - \mathrm{MO}_p^{-p}(k)\right)/p, & \text{if } p < 1/\xi,\ p \neq 0, \\ \ln \mathrm{MO}_0(k) = \mathrm{H}(k), & \text{if } p = 0. \end{cases} \quad (14)$$

Recently, GMs have been used in the semi-parametric estimation of different parameters of rare events, like the *value-at-risk* (see [41], among others) and the *Weibul tail coefficient* (see [17, 19]). Also, the non-regular framework ($p \ge 1/(2\xi)$) has been under development (see [54, 55], among others).

5.2 *Lehmer's Mean-of-Order-p (L_p) EVI-Estimators*

Beyond the average, the p-moments of log-excesses, i.e.

$$M_{k,n}^{(p)} := \frac{1}{k}\sum_{i=1}^{k} \left\{ \ln X_{n-i+1:n} - \ln X_{n-k:n} \right\}^p, \ p \ge 1, \ \left[M_{k,n}^{(1)} \equiv \mathrm{H}(k)\right] \quad (15)$$

introduced in [25], have also played a relevant role in the EVI-estimation, and can more generally be parameterized in $p \in \mathbb{R} \setminus \{0\}$. And a simple generalization of the average is *Lehmer's mean-of-order-p* (see [75]): Given a set of positive numbers $\underline{a} = (a_1, \ldots, a_k)$, such a mean generalizes both the arithmetic mean ($p = 1$) and the harmonic mean ($p = 0$), being defined as

$$L_p(\underline{a}) := \sum_{i=1}^{k} a_i^p / \sum_{i=1}^{k} a_i^{p-1}, \quad p \in \mathbb{R}.$$

The H EVI-estimators can thus be considered as the Lehmer mean-of-order-1 of the k log-excesses $\underline{V} := (V_{ik},\ 1 \le i \le k)$, defined in (5). Following [87] (see also, [66, 88, 89], among others), and with E denoting a standard exponential RV, we can say that $V_{ik} = \xi E_{k-i+1:k} + o_p(1)$. Since $\mathbb{E}(E^p) = \Gamma(p+1)$, $\forall p > -1$, with $\Gamma(\cdot)$ denoting the Gamma function, the law of large numbers enables us to say that

$$\frac{1}{k}\sum_{i=1}^{k} V_{ik}^p \underset{n\to\infty}{\overset{\mathbb{P}}{\longrightarrow}} \Gamma(p+1)\xi^p.$$

Hence the reason for the class of L_p EVI-estimators,

$$\mathrm{L}_p(k) = \mathrm{L}_p(k; \underline{X}_n) := \frac{L_p(\underline{V})}{p} = \frac{1}{p}\,\frac{\sum_{i=1}^{k} V_{ik}^p}{\sum_{i=1}^{k} V_{ik}^{p-1}} = \frac{M_{k,n}^{(p)}}{p M_{k,n}^{(p-1)}} \quad \big[\mathrm{L}_1(k) \equiv \mathrm{H}(k)\big], \tag{16}$$

consistent for all $\xi \ge 0$ and real $p > 0$, and with $M_{k,n}^{(p)}$ defined in (15).

5.3 Asymptotic Comparison at Optimal Levels

At optimal levels, in the sense of minimal asymptotic *root* MSE (RMSE), both L_{OPT}, the optimal L_p, with L_p given in (16), and H_{OPT}, the optimal H_p, with H_p given in (14), outperform the H EVI-estimator, either for IID or even for a wide variety of dependent frameworks. Moreover, and of high theoretical and practical relevance, L_{OPT}, the optimal L_p, beats H_{OPT}, the optimal H_p, a point that strongly favors this new optimal L_p class of EVI-estimators, not yet included in the illustration with simulated data provided in Sect. 5.5. Regarding this topic, further details can be found in [87, 88], already mentioned, and in [13].

5.4 PORT-GM EVI-Estimators

The classes of GM EVI-estimators, either in (14) or in (16), depend on this *tuning* parameter $p \in \mathbb{R}$, are highly flexible, but, as often desired, they are not location-invariant, depending strongly on possible shifts in the model underlying the data, contrarily to what happens to the EVI, which is independent of shifts to the data. It is thus sensible to suggest the use of the classes of PORT-GM EVI-estimators. They are similar in spirit to the PORT-H EVI-estimators, studied in [1], and further considered in [49]. Classes of PORT estimators are based on a *sample of excesses* over a random threshold $X_{n_s:n}$, $n_s := \lfloor ns \rfloor + 1, 0 \le s < 1$,

$$\underline{X}_n^{(s)} := \big(X_{n:n} - X_{\lfloor ns \rfloor+1:n}, \dots, X_{\lfloor ns \rfloor+2:n} - X_{\lfloor ns \rfloor+1:n}\big). \tag{17}$$

For $0 \le s < 1$ and $k < n - n_s$, the PORT-GM class of EVI-estimators has the same functional form of the GM class of EVI-estimators, but with $\underline{X}_n = (X_1, \dots, X_n)$ replaced by the sample of excesses $\underline{X}_n^{(s)}$, in (17). They are thus given by

$$\mathrm{GM}_p(k, s) := \mathrm{GM}_p\big(k; \underline{X}_n^{(s)}\big), \quad [\mathrm{GM} = \mathrm{H} \text{ and } \mathrm{GM} = \mathrm{L}].$$

The PORT H_p have been introduced and studied both theoretically and for finite samples in [39, 53], being the study of the PORT L_p EVI-estimators still under development.

These EVI-estimators are now invariant for both changes of scale and location in the data, and depend on the extra *tuning parameter* s, which provides a highly flexible class of EVI-estimators. Indeed, as shown in [39], for the PORT H_p EVI-estimation, these estimators may compare favorably with the PORT versions of the second-order MVRB EVI-estimators introduced in [18] and already defined in (7), provided that we adequately choose the vector, (p, s), of tuning parameters. Note that the PORT methodology will lead to no change in the asymptotic variance. There is only a change in the asymptotic bias, comparatively with the associated GM EVI-estimators.

5.5 A Short Illustration with Simulated Data

Apart from the H_p EVI-estimators, in (14), we also consider the simplest class of second-order MVRB EVI-estimators, already defined in (7), and, more generally, the class of H_p RB EVI-estimators, introduced in [39], with the functional form,

$$\overline{\mathrm{H}}_p(k) = \mathrm{H}_p(k)\Big(1 - \frac{\hat\beta(1 - p\mathrm{H}_p(k))}{1 - \hat\rho - p\mathrm{H}_p(k)} \Big(\frac{n}{k}\Big)^{\hat\rho}\Big),$$

being $(\hat\beta, \hat\rho)$ an adequate estimators of the vector (β, ρ) of second-order parameters, in the Hall-Welsh class of models, defined in (4). For the estimation of (β, ρ), the algorithm in [68] was followed (see again [45], among others, where additional algorithms are provided).

Just as an illustration of large-scale Monte-Carlo simulation studies undertaken in several articles on the topic (see Gomes et al. [39], among others), we picture Fig. 7, where, for a GEV underlying model, with $\xi = 0.1$, the patterns of simulated expected value (E) and RMSE of the H EVI-estimators, the estimated optimal H_p EVI-estimators, denoted by $\widehat{\mathrm{H}}^*$, and associated RB EVI-estimators, denoted by $\widehat{\overline{\mathrm{H}}}^*$, are presented.

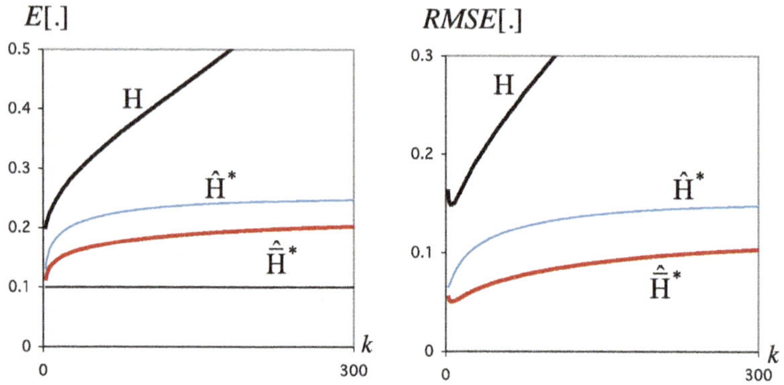

Fig. 7 IID set-up and $\mathrm{GEV}_{0.1}$ underlying model

Similar patterns were obtained for other simulated models.

For all k, there is a reduction in RMSE, as well as in bias, and $\widehat{\mathrm{H}}^*$, the estimated $\mathrm{H}_{\mathrm{OPT}}$ beats the H EVI-estimators for all simulated parents. Even better results have been obtained for the PORT-H and PORT $\widehat{\mathrm{H}}^*$ EVI-estimators (see [39]), and are also expected for $\widehat{\mathrm{L}}^*$ and PORT-$\widehat{\mathrm{L}}^*$ EVI-estimators. As expected, $\widehat{\overline{\mathrm{H}}}^*$ beats $\widehat{\mathrm{H}}^*$ for all k. The choice of the tuning parameters (k, p), or (k, p, s), can be done through heuristic sample-path stability algorithms, like the ones in [52] (see also, [84]). For further algorithmic details on the technique, see also [15, 42], where R-scripts are provided.

6 Overall Comments and Open Research Topics

— Resampling methodologies, like the Jackknife, the GJ and the Bootstrap, have been fruitfully used in the estimation of the EVI, of the extremal index and of the value at risk, but can also be used in the reliable estimation of many other parameters of extreme events, either in an univariate or multivariate framework, a topic still under development in what regards the estimation of the conditional tail expectation, the tail dependence coefficient and the Weibull tail coefficient.

— We have here considered the univariate case only, but EVT is of high relevance both in the multivariate and in the spatial set-ups, whenever dealing with the modeling of extreme events or equivalently the modeling of risk. But indeed, even statistics of univariate extremes is still a quite lively topic of research, of high importance in risk assessment.
— Important developments have appeared recently in the area of spatial extremes, where parametric models, both asymptotic and pre-asymptotic, became again quite relevant. And in a semi-parametric framework, topics like threshold selection and the PORT methodology are still quite challenging, being the PORT L_p EVI-estimation still an open topic of research, as indicated in Sect. 5.
— It is obvious that GMs, often dependent on an extra parameter $p \in \mathbb{R}$, can be used, surely with success, in the estimation of any parameter of extreme events. And alternative reliable EVI-estimators based on GMs (see [86], and references therein) can also play a relevant role in the estimation of those parameters. Indeed, other GMs EVI-estimators, which have been successfully used in the estimation of the Weibull tail coefficient in a research still under development, can be found in pioneer papers, like [56, 57, 94].
— Algorithmic details on the choice of tuning parameters under play are in most cases still under progress, being also sensible to use Bootstrap algorithms of the type of the ones in [46, 63], among others.
— Moreover, the lack of efficiency of the H_p EVI-estimators for $p < 0$ together with the results in [96], related to the robustness of the MO_{-1} EVI-estimators, deserves a further discussion of the topic 'robustness vs efficiency'.
— For some of the GMs estimators, non-regular frameworks, where a non-degenerate non-normal asymptotic behaviour can appear, deserve further attention. The challenge is not to go beyond the value of p that leads to consistency. Several algorithms have already been conceived and are able to partially perform such a goal, but this is a topic still under investigation, being part of an interesting 'tale of the tails'.

Acknowledgements This research was partially funded by National Funds through **FCT**—Fundação para a Ciência e a Tecnologia, Portugal, projects UID/00006/2025, UIDB/00006/2020 (https://doi.org/10.54499/UIDB/00006/2020), UIDP/00006/2020 (CEAUL), and HiTEc Cost Action CA21163.

References

1. Araújo Santos, P., Fraga Alves, M.I., Gomes, M.I.: Peaks over random threshold methodology for tail index and high quantile estimation. REVSTAT–Statist. J. **4**(3), 227–247 (2006). https://doi.org/10.57805/revstat.v4i3.37
2. Balkema, A.A., de Haan, L.: Residual life time at great age. Ann. Probab. **2**, 792–804 (1974). https://doi.org/10.1214/aop/1176996548

3. Beirlant, J., Caeiro, F., Gomes, M.I.: An overview and open research topics in statistics of univariate extremes. REVSTAT–Statist. J. **10**(1), 1–31 (2012). https://doi.org/10.57805/revstat.v10i1.109
4. Beirlant, J., Dierckx, G., Goegebeur, Y., Matthys, G.: Tail index estimation and an exponential regression model. Extremes **2**(2), 177–200 (1999). https://doi.org/10.1023/A:1009975020370
5. Beirlant, J., Figueiredo, F., Gomes, M.I., Vandewalle, B.: Improved reduced-bias tail index and quantile estimators. J. Statist. Plann. Inference **138**(6), 1851–1870 (2008). https://doi.org/10.1016/j.jspi.2007.07.015
6. Beirlant, J., Goegebeur, Y., Segers, J., Teugels, J.: Statistics of Extremes. Theory and Applications. Wiley Series in Probability and Statistics. Wiley, Hoboken (2004). https://doi.org/10.1002/0470012382
7. Beran, J., Schell, D., Stehlík, M.: The harmonic moment tail index estimator: asymptotic distribution and robustness. Ann. Inst. Statist. Math. **66**(1), 193–220 (2014). https://doi.org/10.1007/s10463-013-0412-2
8. Bingham, N.H., Goldie, C.M., Teugels, J.L.: Regular Variation, Encyclopedia of Mathematics and its Applications, vol. 27. Cambridge University Press, Cambridge (1987)
9. Box, G.E.P., Draper, N.R.: Empirical Model-Building and Response Surfaces. Wiley Series in Probability and Mathematical Statistics: Applied Probability and Statistics. Wiley, New York (1987)
10. Brilhante, M.F., Gomes, M.I., Mendonça, S., Pestana, D., Pestana, P.: Generalized beta models and population growth: so many routes to chaos. Fractal Fract. **7**(2) (2023). https://www.mdpi.com/2504-3110/7/2/194
11. Brilhante, M.F., Gomes, M.I., Pestana, D.: A simple generalisation of the Hill estimator. Comput. Statist. Data Anal. **57**, 518–535 (2013). https://doi.org/10.1016/j.csda.2012.07.019
12. Brilhante, M.F., Gomes, M.I., Pestana, D.: The MOP EVI–estimator revisited. In: Pacheco, A., et al. (eds.) New advances in statistical modeling and applications. Studies in Theoretical and Applied Statistics. Sel. Papers Statistics in Society, pp. 163–175. Springer, Cham (2014). https://doi.org/10.1007/978-3-319-05323-3_16
13. Cabral, I., Caeiro, F., Gomes, M.I.: On the comparison of several classical estimators of the extreme value index. Comm. Statist. Theory Methods **51**(1), 179–196 (2022). https://doi.org/10.1080/03610926.2020.1746970
14. Caeiro, F., Gomes, M.I.: On the bootstrap methodology for the estimation of the tail sample fraction. In: Gilli, M., Gonzalez-Rodriguez, G., Nieto-Reyes, A. (eds.) Proceedings of COMPSTAT 2014 — 21st International Conference on Computational Statistics, pp. 545–552. The International Statistical Institute, The Hague (2014)
15. Caeiro, F., Gomes, M.I.: Threshold Selection in Extreme Value Analysis. In: Dey, D., Yan, J. (eds.) Extreme Value Modeling and Risk Analysis, pp. 69–86. CRC Press, Boca Raton, FL (2016)
16. Caeiro, F., Gomes, M.I., Beirlant, J., de Wet, T.: Mean-of-order p reduced-bias extreme value index estimation under a third-order framework. Extremes **19**(4), 561–589 (2016). https://doi.org/10.1007/s10687-016-0261-5
17. Caeiro, F., Gomes, M.I., Henriques-Rodrigues, L.: Estimation of the Weibull tail coefficient through the power mean-of-order-p. In: Bispo, R., Henriques-Rodrigues, L., Alpizar-Jara, R., de Carvalho, M. (eds.) Recent Developments in Statistics and Data Science, pp. 41–53. Springer International Publishing, Cham (2022). https://doi.org/10.1007/978-3-031-12766-3_4
18. Caeiro, F., Gomes, M.I., Pestana, D.: Direct reduction of bias of the classical Hill estimator. REVSTAT–Statist. J. **3**(2), 113–136 (2005). https://doi.org/10.57805/revstat.v3i2.21
19. Caeiro, F., Henriques-Rodrigues, L., Gomes, M.I.: The use of generalized means in the estimation of the Weibull tail coefficient. Comput. Math. Methods pp. Art. ID 7290822, 12 (2022). https://doi.org/10.1155/2022/7290822
20. de Carvalho, M., Pereira, S., Pereira, P., de Zea Bermudez, P.: An extreme value Bayesian Lasso for the conditional left and right tails. J. Agric. Biol. Environ. Stat. **27**(2), 222–239 (2022). https://doi.org/10.1007/s13253-021-00469-9

21. de Carvalho, M., Ramos, A.: Bivariate extreme statistics, II. REVSTAT–Statist. J. **10**(1), 83–107 (2012). https://doi.org/10.57805/revstat.v10i1.112
22. Castillo, E., Hadi, A.S., Balakrishnan, N., Sarabia, J.M.: Extreme Value and Related Models with Applications in Engineering and Science. Wiley Series in Probability and Statistics. Wiley-Interscience, Wiley, Hoboken, NJ (2005)
23. Coles, S.: An Introduction to Statistical Modeling of Extreme Values. Springer Series in Statistics. Springer, London, Ltd., London (2001). https://doi.org/10.1007/978-1-4471-3675-0
24. Davison, A.C., Huser, R.: Statistics of extremes. Ann. Rev. Stat. Appl. **2**, 203–235 (2015). https://doi.org/10.1146/annurev-statistics-010814-020133
25. Dekkers, A.L.M., Einmahl, J.H.J., de Haan, L.: A moment estimator for the index of an extreme-value distribution. Ann. Statist. **17**(4), 1833–1855 (1989). https://doi.org/10.1214/aos/1176347397
26. Efron, B.: Bootstrap methods: another look at the jackknife. Ann. Stat. **7**, 1–26 (1979). https://doi.org/10.1214/aos/1176344552
27. Feuerverger, A., Hall, P.: Estimating a tail exponent by modelling departure from a Pareto distribution. Ann. Statist. **27**(2), 760–781 (1999). https://doi.org/10.1214/aos/1018031215
28. Fisher, R.A., Tippett, L.H.C.: Limiting forms of the frequency distribution of the largest or smallest member of a sample. Proc. Camb. Philos. Soc. **24**, 180–190 (1928). https://doi.org/10.1017/S0305004100015681
29. Fraga Alves, M.I., Gomes, M.I., de Haan, L.: A new class of semi-parametric estimators of the second order parameter. Port. Math. (N.S.) **60**(2), 193–213 (2003). https://eudml.org/doc/50867
30. Gnedenko, B.: Sur la distribution limite du terme maximum d'une série aléatoire. Ann. Math. **44**(3), 423–453 (1943). https://doi.org/10.2307/1968974
31. Gomes, M.I.: Some Probabilistic and Statistical Problems in Extreme Value Theory. Ph.D. thesis, University of Sheffield (1978)
32. Gomes, M.I.: An i-dimensional limiting distribution function of largest values and its relevance to the statistical theory of extremes. In: Taillie, C., Patil, G.P., Baldessari, B.A. (eds.) Statistical distributions in scientific work, vol. 6 (Trieste, 1980), NATO Advanced Study Institute Series C: Mathematical and Physical Sciences, vol. 79, pp. 389–410. Reidel, Dordrecht-Boston, Mass (1981)
33. Gomes, M.I.: Metodologias Jackknife e Bootstrap em Estatística de Extremos. In: Mendes-Lopes, N. et al. (eds.) Actas da II Conferência Anual da S.P.E., pp. 31–46. SPE (in portuguese), Luso, Portugal (1995)
34. Gomes, M.I.: The jackknife and the bootstrap methodologies in the estimation of parameters of rare events. Revista de Estatística **99**, 5–23 (1999)
35. Gomes, M.I.: Extreme values and resampling techniques. In: Temido, G., Ferreira, H. (eds.) Extreme Values and Resampling Techniques, pp. 1–12. Universidade de Coimbra (2002)
36. Gomes, M.I.: Optimal sample fraction, bias reduction in the semi-parametric estimation of rare events' parameters and "related" topics. In: Temido, G., Ferreira, H. (eds.) Extreme Values and Resampling Techniques, pp. 13–18. Universidade de Coimbra (2002)
37. Gomes, M.I.: Resampling methodologies in the field of statistics of univariate extremes. In: Pacheco, A., et al. (eds.) New advances in statistical modeling and applications. Selected Papers Based on the Presentations at the 19th Annual Congress of the Portuguese Statistical Society, SPE 2011, September 28–October 1, 2011, pp. 29–40. Springer, Cham (2014). https://doi.org/10.1007/978-3-319-05323-3_3
38. Gomes, M.I.: Tail dependence coefficient semi-parametric estimation through a generalized mean. In: Oliveira, A., et al (eds.) Book of Abstracts, pp. 14–16. Universidade Aberta editions (2019)
39. Gomes, M.I., Brilhante, M.F., Pestana, D.: New reduced-bias estimators of a positive extreme value index. Comm. Statist. Simulation Comput. **45**(3), 833–862 (2016). https://doi.org/10.1080/03610918.2013.875567

40. Gomes, M.I., Caeiro, F.: Efficiency of partially reduced-bias mean-of-order-p versus minimum-variance reduced-bias extreme value index estimation. In: Gilli, M., et al. (eds.) Proceedings of COMPSTAT 2014 — 21st International Conference on Computational Statistics, pp. 289–298. The International Statistical Institute, The Hague (2014)
41. Gomes, M.I., Caeiro, F., Figueiredo, F., Henriques-Rodrigues, L., Pestana, D.: Reduced-bias and partially reduced-bias mean-of-order-p value-at-risk estimation: a Monte-Carlo comparison and an application. J. Stat. Comput. Simul. **90**(10), 1735–1752 (2020). https://doi.org/10.1080/00949655.2020.1746787
42. Gomes, M.I., Caeiro, F., Henriques-Rodrigues, L., Manjunath, B.G.: Bootstrap methods in statistics of extremes. In: Longin, F. (ed.) Extreme Events in Finance: A Handbook of Extreme Value Theory and its Applications, Chap. 6, pp. 117–138. Wiley (2016). https://onlinelibrary.wiley.com/doi/abs/10.1002/9781118650318.ch6
43. Gomes, M.I., Canto e Castro, L., Fraga Alves, M.I., Pestana, D.: Statistics of extremes for IID data and breakthroughs in the estimation of the extreme value index: Laurens de Haan leading contributions. Extremes **11**(1), 3–34 (2008). https://doi.org/10.1007/s10687-007-0048-9
44. Gomes, M.I., Figueiredo, F.: Bias reduction in risk modelling: semi-parametric quantile estimation. Test **15**(2), 375–396 (2006). https://doi.org/10.1007/BF02607058
45. Gomes, M.I., Figueiredo, F., Martins, M.J., Neves, M.M.: Resampling methodologies and reliable tail estimation. South African Statist. J. **49**(1), 1–20 (2015)
46. Gomes, M.I., Figueiredo, F., Neves, M.M.: Adaptive estimation of heavy right tails: resampling-based methods in action. Extremes **15**(4), 463–489 (2012). https://doi.org/10.1007/s10687-011-0146-6
47. Gomes, M.I., Figueiredo, F., Neves, M.M.: Adaptive choice of thresholds and the methodology: an empirical study. In: da Silva, J.L., Caeiro, F., Natário, I., Braumann, C.A. (eds.) Advances in Regression, Survival Analysis, Extreme Values, Markov Processes and Other Statistical Applications, Studies in Theoretical and Applied Statistics, Selected Papers of the Statistical Societies, pp. 203–211. Springer, Heidelberg (2013). https://doi.org/10.1007/978-3-642-34904-1_21
48. Gomes, M.I., Fraga Alves, M.I., Neves, C.: Análise de Valores Extremos: uma Introdução. Edições SPE & INE (2013)
49. Gomes, M.I., Fraga Alves, M.I., Santos, P.A.: PORT Hill and moment estimators for heavy-tailed models. Comm. Statist. Simulation Comput. **37**(6–7), 1281–1306 (2008). https://doi.org/10.1080/03610910802050910
50. Gomes, M.I., Guillou, A.: Extreme value theory and statistics of univariate extremes: a review. Int. Stat. Rev. **83**(2), 263–292 (2015). https://doi.org/10.1111/insr.12058
51. Gomes, M.I., Henriques-Rodrigues, L., Figueiredo, F.: Resampling-based methodologies in statistics of extremes: Environmental and financial applications. In: Bourguignon, J.P., Jeltsch, R., Pinto, A.A., Viana, M. (eds.) Mathematics of Energy and Climate Change, pp. 63–181. CIM Series in Mathematical Sciences, Springer International Publishing Switzerland (2015). https://doi.org/10.1007/978-3-319-16121-1_6
52. Gomes, M.I., Henriques-Rodrigues, L., Fraga Alves, M.I., Manjunath, B.G.: Adaptive PORT-MVRB estimation: an empirical comparison of two heuristic algorithms. J. Stat. Comput. Simul. **83**(6), 1129–1144 (2013). https://doi.org/10.1080/00949655.2011.652113
53. Gomes, M.I., Henriques-Rodrigues, L., Manjunath, B.G.: Mean-of-order-p location-invariant extreme value index estimation. REVSTAT–Statist. J. **14**(3), 273–296 (2016). https://doi.org/10.57805/revstat.v14i3.190
54. Gomes, M.I., Henriques-Rodrigues, L., Pestana, D.: A generalized mean under a non-regular framework and extreme value index estimation. In: Zafeiris, K., et al. (eds.) Data Analysis and Related Applications 2, Chap. 16, pp. 237–250. Wiley (2022). https://onlinelibrary.wiley.com/doi/abs/10.1002/9781394165544.ch16
55. Gomes, M.I., Henriques-Rodrigues, L., Pestana, D.: Non-regular frameworks and the mean-of-order p extreme value index estimation. J. Stat. Theory Pract. **16**(3), Paper No. 37, 17 (2022). https://doi.org/10.1007/s42519-022-00264-w

56. Gomes, M.I., Martins, M.J.: Efficient alternatives. to the Hill estimator. In: Gomes, M.I., Pestana, D., Canto e Castro, L., Fraga Alves, M.I., Martins, M.J. (eds.) Extreme Values and Additive Laws: Extended Abstracts, pp. 40–43. Lisbon, Portugal: C.E.A.U.L. editions (1999)
57. Gomes, M.I., Martins, M.J.: Generalizations of the Hill estimator-asymptotic versus finite sample behaviour. J. Statist. Plann. Inference **93**(1–2), 161–180 (2001). https://doi.org/10.1016/S0378-3758(00)00201-9
58. Gomes, M.I., Martins, M.J.: "Asymptotically unbiased" estimators of the tail index based on external estimation of the second order parameter. Extremes **5**(1), 5–31 (2002). https://doi.org/10.1023/A:1020925908039
59. Gomes, M.I., Martins, M.J., Neves, M.: Alternatives to a semi-parametric estimator of parameters of rare events—the jackknife methodology. Extremes **3**(3), 207–229 (2000). https://doi.org/10.1023/A:1011470010228
60. Gomes, M.I., Martins, M.J., Neves, M.M.: Generalized jackknife semi-parametric estimators of the tail index. Port. Math. (N.S.) **59**(4), 393–408 (2002). http://eudml.org/doc/50313
61. Gomes, M.I., Martins, M.J., Neves, M.M.: Improving second order reduced bias extreme value index estimation. REVSTAT–Statist. J. **5**(2), 177–207 (2007). https://doi.org/10.57805/revstat.v5i2.48
62. Gomes, M.I., Martins, M.J., Neves, M.M.: Generalized jackknife-based estimators for univariate extreme-value modeling. Comm. Statist. Theory Methods **42**(7), 1227–1245 (2013). https://doi.org/10.1080/03610926.2012.725263
63. Gomes, M.I., Mendonça, S., Pestana, D.: Adaptive reduced-bias tail index and VaR estimation via the bootstrap methodology. Comm. Statist. Theory Methods **40**(16), 2946–2968 (2011). https://doi.org/10.1080/03610926.2011.562782
64. Gomes, M.I., Miranda, C., Viseu, C.: Reduced-bias tail index estimation and the Jackknife methodology. Statist. Neerlandica **61**(2), 243–270 (2007). https://doi.org/10.1111/j.1467-9574.2007.00346.x
65. Gomes, M.I., Oliveira, O.: The bootstrap methodology in statistics of extremes—choice of optimal sample fraction. Extremes **4**(4), 331–358 (2001). https://doi.org/10.1023/A:1016592028871
66. Gomes, M.I., Penalva, H., Caeiro, F., Neves, M.M.: Non-reduced versus reduced-bias estimators of the extreme value index—efficiency and robustness. In: Colubi, A., et al. (eds.) Proceedings of COMPSTAT 2016—22nd International Conference on Computational Statistics, pp. 279–290. The International Statistical Institute, The Hague (2016)
67. Gomes, M.I., Pereira, H., Miranda, M.C.: Revisiting the role of the jackknife methodology in the estimation of a positive tail index. Comm. Statist. Theory Methods **34**(2), 319–335 (2005). https://doi.org/10.1080/03610920509342423
68. Gomes, M.I., Pestana, D.: A sturdy reduced-bias extreme quantile (VaR) estimator. J. Am. Statist. Assoc. **102**(477), 280–292 (2007). https://doi.org/10.1198/016214506000000799
69. Gomes, M.I., Pestana, D.: Corrections: A sturdy reduced-bias extreme quantile (VaR) estimator. J. Amer. Statist. Assoc. **104**(486), 872 (2009). https://doi.org/10.1198/jasa.2009.0140
70. Gomes, M.I., Reiss, R.D., Thomas, M.: An Overview of Reduced-Bias Estimation. In: Reiss, R.-D., Thomas, M. (eds.) Statistical Analysis of Extreme Values with Applications to Insurance, Finance, Hydrology and Other Fields, 3rd edn., Chap. 6, pp. 190–206. Birkhäuser Verlag, Basel (2007)
71. Gray, H.L., Schucany, W.R.: The Generalized Jackknife Statistic, Statistics: Textbooks and Monographs, vol. 1. Marcel Dekker Inc, New York (1972)
72. Gumbel, E.J.: Statistics of Extremes. Columbia University Press, New York (1958)
73. Hall, P.: On some simple estimates of an exponent of regular variation. J. Roy. Statist. Soc. Ser. B **44**(1), 37–42 (1982). https://www.jstor.org/stable/2984706
74. Hall, P., Welsh, A.H.: Adaptive estimates of parameters of regular variation. Ann. Statist. **13**(1), 331–341 (1985). https://doi.org/10.1214/aos/1176346596
75. Havil, J.: Gamma: Exploring Euler's Constant. Princeton University Press, Princeton, NJ (2003)
76. Henriques-Rodrigues, L., Gomes, M.I., Fraga Alves, M.I., Neves, C.: PORT-estimation of a shape second-order parameter. REVSTAT–Statist. J. **12**(3), 299–328 (2014). https://doi.org/10.57805/revstat.v12i3.155

77. Henriques-Rodrigues, L., Gomes, M.I., Manjunath, B.G.: Estimation of a scale second-order parameter related to the PORT methodology. J. Stat. Theory Pract. **9**(3), 571–599 (2015). https://doi.org/10.1080/15598608.2014.969816
78. Hill, B.M.: A simple general approach to inference about the tail of a distribution. Ann. Statist. **3**(5), 1163–1174 (1975). https://www.jstor.org/stable/2958370
79. Hüsler, J., Peng, L.: Review of testing issues in extremes: in honor of Professor Laurens de Haan. Extremes **11**(1), 99–111 (2008). https://doi.org/10.1007/s10687-007-0052-0
80. Israelsson, J., Black, E., Neves, C., Walshaw, D.: Estimation and reduced bias estimation of the residual dependence index with unnamed marginals (2021). arXiv:2112.02899v2
81. Markovich, N.: Nonparametric Analysis of Univariate Heavy-tailed Data. Wiley Series in Probability and Statistics. Wiley, Chichester (2007). https://doi.org/10.1002/9780470723609
82. Naveau, P., Huser, R., Ribereau, P., Hannart, A.: Modeling jointly low, moderate, and heavy rainfall intensities without a threshold selection. Water Resour. Res. **52**(4), 2753–2769 (2016). https://doi.org/10.1002/2015WR018552
83. Neves, C., Fraga Alves, M.I.: Testing extreme value conditions—an overview and recent approaches. REVSTAT–Statist. J. **6**(1), 83–100 (2008). https://doi.org/10.57805/revstat.v6i1.59
84. Neves, M.M., Gomes, M.I., Figueiredo, F., Prata Gomes, D.: Modeling extreme events: sample fraction adaptive choice in parameter estimation. J. Stat. Theory Pract. **9**(1), 184–199 (2015). https://doi.org/10.1080/15598608.2014.890984
85. Paulauskas, V., Vaičiulis, M.: On an improvement of Hill and some other estimators. Lith. Math. J. **53**(3), 336–355 (2013). https://doi.org/10.1007/s10986-013-9212-x
86. Paulauskas, V., Vaičiulis, M.: A class of new tail index estimators. Ann. Inst. Stat. Math. **69**(2), 461–487 (2017). https://doi.org/10.1007/s10463-015-0548-3
87. Penalva, H., Caeiro, F., Gomes, M.I., Neves, M.M.: An efficient naive generalization of the Hill estimator—discrepancy between asymptotic and finite sample behaviour. Notas e Comunicações CEAUL 02/2016 (2016)
88. Penalva, H., Gomes, M.I., Caeiro, F., Neves, M.M.: A couple of non reduced bias generalized means in extreme value theory: an asymptotic comparison. REVSTAT–Statist. J. **18**(3), 281–298 (2020). https://doi.org/10.57805/revstat.v18i3.301
89. Penalva, H., Gomes, M.I., Caeiro, F., Neves, M.M.: Lehmer's mean-of-order-p extreme value index estimation: a simulation study and applications. J. Appl. Stat. **47**(13-15), 2825–2845 (2020). https://doi.org/10.1080/02664763.2019.1694871
90. Peng, L.: Asymptotically unbiased estimators for the extreme-value index. Stat. Probab. Lett. **38**(2), 107–115 (1998). https://doi.org/10.1016/S0167-7152(97)00160-0
91. Pickands III, J.: Statistical inference using extreme order statistics. Ann. Stat. **3**, 119–131 (1975). https://www.jstor.org/stable/2958083
92. Reiss, R.D., Thomas, M.: Statistical Analysis of Extreme Values with Applications to Insurance, Finance, Hydrology and Other Fields, 2nd, 3rd edn. Birkhäuser Verlag, Basel (2001, 2007)
93. Scarrott, C., MacDonald, A.: A review of extreme value threshold estimation and uncertainty quantification. REVSTAT–Statist. J. **10**(1), 33–60 (2012). https://doi.org/10.57805/revstat.v10i1.110
94. Segers, J.: Residual estimators. J. Stat. Plann. Inference **98**(1–2), 15–27 (2001). https://doi.org/10.1016/S0378-3758(00)00321-9
95. Smith, R.L.: Estimating tails of probability distributions. Ann. Stat. **15**(3), 1174–1207 (1987). https://doi.org/10.1214/aos/1176350499
96. Stehlík, M., Potocký, R., Waldl, H., Fabián, Z.: On the favorable estimation for fitting heavy tailed data. Comput. Stat. **25**(3), 485–503 (2010). https://doi.org/10.1007/s00180-010-0189-1
97. The Editors of Encyclopedia: North Sea flood. Encyclopedia Britannica. https://www.britannica.com/event/North-Sea-flood

98. Vandewalle, B., Beirlant, J.: On univariate extreme value statistics and the estimation of reinsurance premiums. Insur. Math. Econ. **38**(3), 441–459 (2006). https://doi.org/10.1016/j.insmatheco.2005.11.002
99. Weissman, I.: Estimation of parameters and large quantiles based on the k largest observations. J. Am. Stat. Assoc. **73**(364), 812–815 (1978). https://doi.org/10.2307/2286285

Multiplicity, Stability and Symmetry of Periodic Orbits of Reeb Flows

Leonardo Macarini

Abstract Reeb flows form a prominent class of Hamiltonian systems on regular energy levels. Since these flows have no singularities, the most basic dynamical objects are the periodic orbits. In this paper, we survey some results on the multiplicity, stability and symmetry of periodic orbits of Reeb flows without assuming generic conditions.

1 Introduction

Let (M^{2n+1}, ξ) be a closed co-oriented contact manifold. Let α be a contact form supporting ξ (i.e. such that $\ker \alpha = \xi$) and denote by R_α the corresponding Reeb vector field uniquely characterized by the equations $\iota_{R_\alpha} d\alpha = 0$ and $\alpha(R_\alpha) = 1$. Reeb flows form a prominent class of Hamiltonian systems on regular energy levels. Indeed, if M is a contact type hypersurface in a symplectic manifold W and $H : W \to \mathbb{R}$ is a Hamiltonian such that M is a regular energy level of H, then the Hamiltonian flow of H on M is a reparametrization of the Reeb flow. There are several important examples, including proper homogeneous Hamiltonians $H : \mathbb{R}^{2n} \to \mathbb{R}$ and geodesic flows.

We want to study the dynamics of Reeb flows. Since these flows have no singularities, the most basic dynamical objects are the periodic orbits. So, we want to study the periodic orbits of Reeb flows. Under generic conditions on the contact form α (i.e. when α belongs to a residual subset of contact forms in a suitable topology), several results have been established; see, for instance, [43, 68, 72] and references therein. Although the generic case is very interesting and important, the general case is much harder and, throughout this paper, we will consider this case, *not* assuming generic conditions, although we may assume some *explicit* conditions that hold generically (e.g. non-degeneracy).

L. Macarini—(on leave from Instituto Superior Técnico, University of Lisbon).

L. Macarini (✉)
IMPA, Rio de Janeiro, Brazil
e-mail: leonardo@impa.br

M. Abreu et al. (eds.), *Brazil-Portugal Mathematics*, Coimbra Mathematical Texts 4,
https://doi.org/10.1007/978-3-031-98455-6_7

The study of periodic orbits of general flows, without generic conditions, can be something completely out of reach. However, for Reeb flows we have a variational structure that allows us to consider Floer type homologies associated to the corresponding action functional. It is a key tool in the recent results about periodic Reeb orbits and therefore play a key role in most of the results discussed in this survey.

Several questions can be asked about periodic Reeb orbits, but in this survey we will focus on three basic ones. Firstly, the classical question on the *multiplicity* of periodic orbits, that is, the number of geometrically distinct closed Reeb orbits. Secondly, the also classical question on the *stability* of periodic orbits, that is, the existence of *elliptic* closed Reeb orbits. Thirdly, we will address the problem of the *symmetry* of periodic orbits, that is, the existence and multiplicity of symmetric closed Reeb orbits whenever the contact form is invariant under a suitable group action. These three questions are considered in Sects. 2, 3 and 4 respectively.

2 Multiplicity of Periodic Orbits for Reeb Flows

The problem of the existence of periodic orbits on energy levels of Hamiltonian flows is a long-standing question in Hamiltonian Dynamics. This problem traces back to Poincaré and had a strong revival after the pioneering work of Rabinowitz [73–75] who proved the existence of at least one periodic orbit on every starshaped energy hypersurface in $\mathbb{R}^{2n}$.

Rabinowitz's work motivated the introduction of the Weinstein conjecture [88] which asks the existence of a periodic orbit for every Reeb flow on a closed contact manifold and is one of the most fundamental driving questions in Contact Topology. Although this conjecture is open in general, several partial positive results have been achieved, in particular Taubes' celebrated proof in dimension three [81].

A natural problem is what is beyond the Weinstein conjecture, namely, what one can say about the number of geometrically distinct periodic orbits of Reeb flows besides the existence of just one periodic orbit. As pointed out in [44], two complementary classes of contact manifolds can be considered in this problem: those for which the dimension of the contact or symplectic homology groups is *unbounded* and those for which the dimension of these groups is *bounded*. These two cases are discussed in Sects. 2.1 and 2.2 respectively.

2.1 The Homologically Unbounded Case

Let (M, ξ) be a closed contact manifold admitting a strong symplectic filling (W, Ω) such that $c_1(TW)|_{H^2(W,\mathbb{R})} = 0$ and Ω is atoroidal, that is, $\int_{T^2} f^*\Omega = 0$ for every smooth map $f : T^2 \to W$. Then one can associate to W its linearized contact homology $\mathrm{HC}_*(W)$ with rational coefficients and a grading given by the Conley-Zehnder index. It was introduced by Eliashberg, Givental and Hofer in their seminal paper

[41]. We say that (M, ξ) is *homologically unbounded* if there exists such a filling W and a sequence $k_i \to \infty$ such that $\dim \mathrm{HC}_{k_i}(W) \to \infty$ as $i \to \infty$. Otherwise, we say that (M, ξ) is *homologically bounded.*

The following result shows that Reeb flows in homologically unbounded contact manifolds carry infinitely many simple periodic orbits.

Theorem 1 ([58]) *Suppose that (M, ξ) is homologically unbounded. Then the Reeb flow of every contact form supporting ξ has infinitely many geometrically distinct periodic orbits.*

Remark 2 The proof of Theorem 1 relies on the machinery of contact homology which is yet to be made completely rigorous due to transversality issues; cf. [57, 69, 70].

Remark 3 Actually, the result proved in [58] is more general and works for the linearized contact homology associated to a homotopy class of augmentations of the differential graded algebra associated to a non-degenerate contact form supporting ξ. In particular, M can be not fillable.

Examples of homologically unbounded contact manifolds can be obtained by unit cosphere bundles. More precisely, given a closed oriented spin manifold N, it is proved in [26] that

$$\mathrm{HC}_*(D^*N) \cong H_*(\Lambda N/S^1, N; \mathbb{Q}), \tag{1}$$

where D^*N is the unit codisk bundle (that provides a natural filling of the unit cosphere bundle S^*N), ΛN is the free loop space on N and $N \subset \Lambda N$ indicates the subset of constant loops. A result due to Vigué-Poirrier and Sullivan [82] establishes that if N is simply connected then the rank of $H_*(\Lambda N; \mathbb{Q})$ is asymptotically unbounded if, and only if, the rational cohomology algebra of N is not generated by a single class. One can show that the rank of $H_*(\Lambda N; \mathbb{Q})$ is asymptotically unbounded if, and only if, so is the rank of $H_*(\Lambda N/S^1, N; \mathbb{Q})$. Consequently, we have the following generalization of a celebrated result due to Gromoll and Meyer [52]. It was also proved by McLean [66] for more general coefficients using symplectic homology.

Corollary 4 ([58]) *Let N be a closed oriented spin manifold and assume that the rank of $H_*(\Lambda N; \mathbb{Q})$ is asymptotically unbounded. Then every hypersurface in T^*N which is fiberwise starshaped with respect to the zero section has infinitely many, geometrically distinct, periodic orbits. In particular, the result holds if N is simply connected and its cohomology algebra over $\mathbb{Q}$ is not generated by a single class.*

Remark 5 The assumption that N is spin is used only to achieve the isomorphism (1). The same isomorphism should hold without this hypothesis if we take the homology of ΛN with a suitable local system of coefficients [1–3]. This assumption is not needed in [66].

The key ingredient to prove Theorem 1 is the *local contact homology* $\mathrm{HC}_*(\gamma)$ associated to an isolated closed orbit γ, introduced in [58]. The crucial property is a uniform bound for the rank of local contact homology under iterations, established in the following theorem.

Theorem 6 ([58]) *Let γ be a periodic orbit of the Reeb flow such that γ^j is isolated for every $j \in \mathbb{N}$. Then there exists a constant $B > 0$ satisfying* $\dim \mathrm{HC}_l(\gamma^j) < B$ *for every $j \in \mathbb{N}$ and $l \in \mathbb{Z}$.*

Once we have this uniform bound, the rough idea of the proof of Theorem 1 is that finitely many simple periodic orbits cannot generate the total homology if it is homologicaly unbounded. Morse type inequalities are also obtained in [58] relating the local and global homologies.

2.2 The Homologically Bounded Case

When the contact manifold is homologically bounded, the multiplicity of periodic orbits is more involved since, in this context, we do have examples of Reeb flows with finitely many simple periodic orbits. In this situation, the multiplicity of periodic orbits resembles the Hamiltonian Conley conjecture (HCC) for Hamiltonian diffeomorphisms on closed symplectic manifolds where there is also no (Floer) homological growth. The HCC states that every Hamiltonian diffeomorphism carries infinitely many simple periodic points *whenever the symplectic manifold satisfies some natural general requirements*. This conjecture was proved by Salamon and Zehnder [79] for weakly non-degenerate Hamiltonian diffeomorphisms on aspherical symplectic manifolds and by Hingston [55] for general Hamiltonian diffeomorphisms on the standard symplectic torus T^{2n}. These achievements culminate with Ginzburg's celebrated proof [42] of the HCC for general Hamiltonian diffeomorphisms on aspherical symplectic manifolds. However, the HCC easily fails if we drop the assumption of asphericity: consider, for instance, an irrational rotation on S^2 which has only the poles as periodic points.

Similarly, the so called contact Conley conjecture (CCC), introduced in [46], establishes that every Reeb flow on a contact manifold has infinitely many simple closed orbits whenever the contact manifold meets some natural general conditions. With this conjecture in mind, our second group of contact manifolds (homologically bounded) naturally splits in two subgroups: the one given by contact manifolds that satisfy the CCC and the other for which the CCC fails.

In order to study these two subgroups, let us consider a prequantization S^1-bundle (M^{2n+1}, ξ) of a closed integral symplectic manifold (B, ω). This can be considered as a prototypical example of a homologically bounded contact manifold. As a matter of fact, under suitable hypotheses on B, its positive equivariant symplectic homology is given by a sum of copies of the singular homology of the basis with a shift in the degree; see, for instance, [8, 9, 46, 47].

In [46], it is proved (under some mild extra assumptions) that the Reeb flow of every index-admissible contact form on M carries infinitely many simple closed orbits whenever ω vanishes on $\pi_2(B)$. (A non-degenerate contact form α on M^{2n+1} is *index–admissible* if its Reeb flow has no contractible closed orbit with Conley–Zehnder index $2-n$ or $2-n \pm 1$ (this is the natural condition to have a well defined cylindrical contact homology). In general, α is index–admissible when there exists a sequence of non-degenerate index–admissible forms C^1-converging to α.) The hypothesis of index-admissibility was dropped in [47], furnishing the following result. It gives a proof of the CCC for these prequantizations, showing that the first subgroup contains a large class of examples.

Theorem 7 ([47]) *Let (M, ξ) be a prequantization S^1-bundle of (B, ω) and let α be a contact form on M supporting ξ. Assume that*

(i) *ω vanishes on $\pi_2(B)$,*
(ii) *$\pi_1(M)$ is torsion free, and*
(iii) *$c_1(\xi) \in H^2(M; \mathbb{Q})$ is atoroidal.*

Then the Reeb flow of α has infinitely many simple closed orbits with contractible projections to B. Assume furthermore that the Reeb flow has finitely many closed Reeb orbits in the free homotopy class $\mathfrak{f}$ of the fiber and that these orbits are weakly non-degenerate, i.e., at least one of the Floquet multipliers is different from 1. Then for every sufficiently large prime k the Reeb flow of α has a simple closed orbit in the class $\mathfrak{f}^k$.

Remark 8 The proof of Theorem 7 relies on the machinery of contact homology which is yet to be made completely rigorous. A proof in the non-degenerate case based on equivariant symplectic homology is presented in [50]. The equivariant symplectic homology is a symplectic invariant introduced by Viterbo [83] and developed by Bourgeois and Oancea [19–23]. As proved in [23], the positive equivariant symplectic homology is isomorphic to contact homology whenever the late is well defined.

The hypothesis that ω vanishes on $\pi_2(B)$ is crucial. Indeed, it is easy to see that the CCC fails when we drop this assumption, as the example of the irrational ellipsoid shows. More generally, every prequantization of a closed symplectic manifold admitting a Hamiltonian circle action with finitely many fixed points does not satisfy the CCC and all these symplectic manifolds are necessarily not aspherical. Thus, without this assumption we have, in general, to obtain a (ideally sharp) lower bound for the number of simple closed orbits.

Consider, for instance, the case of the standard contact sphere S^{2n+1} which is a prequantization bundle of $\mathbb{C}P^n$. In dimension three, it was proved independently by Cristofaro-Gardiner and Hutchings [29] and Ginzburg, Hein, Hryniewicz and the author [48] that every contact form on S^3 has at least two simple closed orbits; see also [61] where an alternate proof was given using a result from [48]. Unfortunately, the situation in higher dimensions is wide open. As a matter of fact, a long standing conjecture establishes that every contact form α on S^{2n+1} has at least $n+1$ simple closed Reeb orbits. This conjecture, however, is very far from being proved when

$n \geq 2$. Indeed, it is not even known if there is more than one simple closed Reeb orbit if $n \geq 2$.

The situation changes dramatically once we impose index assumptions on the closed orbits, convexity, symmetry or non-degeneracy. Let us discuss some results under these hypotheses.

2.2.1 Multiplicity Results Assuming Index Assumptions

Given a closed Reeb orbit γ we denote by $\mu(\gamma)$ and $\mu^+(\gamma)$ the lower semicontinous and upper semicontinuous extensions of the Conley-Zehnder index respectively; see [8, Sect. 3] for precise definitions. Moreover, given a real number x we let $\lfloor x \rfloor = \sup\{k \in \mathbb{Z};\ k \leq x\}$ and $\lceil x \rceil = \inf\{k \in \mathbb{Z};\ k \geq x\}$.

Note that there is a natural bijection between contact forms α on $(S^{2n+1}, \xi_{\mathrm{std}})$ and starshaped hypersurfaces Σ_α in $\mathbb{R}^{2n+2}$ (recall that a hypersurface in $\mathbb{R}^{2n+2}$ is starshaped if it is closed and transverse to the radial vector field) in the following way. Consider the restriction of the Liouville form λ to the unit sphere $S^{2n+1} \subset \mathbb{R}^{2n+2}$. A contact form α supporting the (cooriented) standard contact structure ξ_{std} is a 1-form given by $f\lambda|_{S^{2n+1}}$, where $f : S^{2n+1} \to \mathbb{R}$ is a positive function. The bijection is given by

$$\alpha = f\lambda|_{S^{2n+1}} \longleftrightarrow \Sigma_\alpha = \{\sqrt{f(x)}x;\ x \in S^{2n+1}\}$$

and it satisfies the property that if Σ_α is the energy level of a homogeneous of degree two Hamiltonian $H_\alpha : \mathbb{R}^{2n+2} \to \mathbb{R}$ then the Hamiltonian flow of H_α on Σ_α is equivalent to the Reeb flow of α. We say that α is *convex* if Σ_α bounds a strictly convex domain.

Let α be a convex contact form on S^{2n+1} with $n \geq 1$. It was proved by Ekeland and Hofer [40] that α has at least two simple closed orbits. A remarkable result due to Long and Zhu [65] improved this lower bound, showing that α has at least $\lfloor (n+1)/2 \rfloor + 1$ simple closed orbits and at least $n+1$ simple closed orbits whenever α is non-degenerate. Later on this result was improved by Wang [87] when n is even, showing the existence of $\lceil (n+1)/2 \rceil + 1$ simple closed orbits. In [86], Wang proved that α has at least four simple closed orbits when $n = 3$.

The importance of convexity in these works is two-fold: it implies that every closed orbit γ of α satisfies $\mu(\gamma) \geq n+2$ and it allows the use of classical variational methods to detect the periodic orbits. The proofs are based mainly on these two ingredients and a combinatorial argument using index theory.

However, convexity is not a symplectic invariant: clearly the image of a convex region by a symplectomorphism does not need to be convex in general. This raises the natural question of how to generalize the previous results replacing the hypothesis of convexity by a *symplectic condition*. In this sense, we have the following important definition introduced by Hofer, Wysocki and Zehnder [56].

Definition 9 A contact form α on the standard contact sphere S^{2n+1} is dynamically convex if every closed closed orbit γ satisfies $\mu(\gamma) \geq n+2$.

It is not hard to see that convexity implies dynamical convexity, see [8, 56, 64]. Clearly, dynamical convexity is invariant under contactomorphisms. Moreover, dynamical convexity is more general than convexity: there are contact forms on the standard contact sphere that are dynamically convex and are not contactomorphic to a convex one [24, 25]; see also [11, 49].

As noticed in [8], the lower bound $n+2$ has an important meaning: *it corresponds to the lowest degree with non-trivial contact homology (or positive equivariant symplectic homology) for the standard contact structure ξ_{std} on S^{2n+1}*. Indeed, an easy computation shows that

$$\mathrm{HC}_*(S^{2n+1}) \cong \begin{cases} \mathbb{Q} & \text{if } * = n+2k \text{ and } k \in \{1,2,3,\dots\} \\ 0 & \text{otherwise.} \end{cases}$$

This observation motivated the following definition, introduced in [8], which generalizes the notion of dynamical convexity for more general contact manifolds. In order to deal with contact manifolds that are not simply connected, we will make use of the fact that cylindrical contact homology has a filtration given by the free homotopy class of the periodic orbits: $\mathrm{HC}_*(M) = \oplus_{a\in[S^1,M]}\mathrm{HC}_*^a(M)$.

Definition 10 ([8]) Let (M,ξ) be a closed contact manifold such that $c_1(\xi)|_{H_2(M;\mathbb{R})} = 0$. Let $k_- = \inf\{k \in \mathbb{Z};\, \mathrm{HC}_k(M) \neq 0\}$ and $k_+ = \sup\{k \in \mathbb{Z};\, \mathrm{HC}_k(M) \neq 0\}$. A contact form α on M is positively (resp. negatively) dynamically convex if k_- is finite and $\mu(\gamma) \geq k_-$ (resp. k_+ is finite and $\mu^+(\gamma) \leq k_+$) for every periodic orbit γ of R_α. Similarly, let a be a free homotopy class in M, $k_-^a = \inf\{k \in \mathbb{Z};\, \mathrm{HC}_k^a(M) \neq 0\}$ and $k_+^a = \sup\{k \in \mathbb{Z};\, \mathrm{HC}_k^a(M) \neq 0\}$. A contact form α is positively (resp. negatively) a-dynamically convex if k_-^a is finite and $\mu(\gamma) \geq k_-^a$ (resp. k_+^a is finite and $\mu^+(\gamma) \leq k_+^a$) for every periodic orbit γ of R_α with free homotopy class a.

Remark 11 This definition can be given in an analogous way using positive equivariant symplectic homology instead of contact homology.

Remark 12 As mentioned before, a convex contact form on the standard contact sphere S^{2n+1} is (positively) dynamically convex. Consider a lens space $L_p^{2n+1}(\ell_0,\ell_1,\dots,\ell_n)$ obtained as a finite quotient of S^{2n+1} with the induced contact structure as discussed in Sect. 4. We say that a contact form α on $L_p^{2n+1}(\ell_0,\ell_1,\dots,\ell_n)$ is convex if so is its lift to S^{2n+1}. It is proved in [11, Theorem 1.4] that if α is convex then it is positively a-dynamically convex for every $a \in \pi_1(L_p^{2n+1}(\ell_0,\ell_1,\dots,\ell_n))$.

Remark 13 The motivation to consider *negative* dynamical convexity comes from applications to magnetic flows; see Sect. 3 and [8, Sect. 2.3].

In [9], we give a lower bound for the number of geometrically distinct contractible periodic orbits of dynamically convex contact forms on prequantizations of symplectic manifolds that are not aspherical. More precisely, we have the following result. In the statement,

$$c_B := \inf\{k > 0;\ \exists S \in \pi_2(B) \text{ such that } \langle c_1(TB), S\rangle = k\}$$

denotes the minimal Chern number of the symplectic manifold B and $\chi(B)$ is its Euler characteristic. In what follows, we will call a contact form α just *dynamically convex* if it is either positively or negatively dynamically convex.

Theorem 14 ([9]) *Let (M^{2n+1}, ξ) be a prequantization S^1-bundle of a closed symplectic manifold (B, ω) such that $c_1(\xi)|_{\pi_2(M)} = 0$, $\omega|_{\pi_2(B)} \neq 0$ and $c_B \geq n$. Let α be a 0-dynamically convex contact form supporting ξ. Then the following assertions hold:*

(A) Suppose that $\chi(B) \neq 0$. Then the Reeb flow of α carries at least two geometrically distinct contractible periodic orbits.
(B) Suppose, if $c_B = n$, that $H_k(B; \mathbb{Q}) = 0$ for every odd k. If α is non-degenerate then its Reeb flow carries at least r_B geometrically distinct contractible periodic orbits, where r_B is the total rank of $H_*(B; \mathbb{Q})$.

Remark 15 The proof of Theorem 14 relies on the machinery of contact homology which is yet to be made completely rigorous. However, when α is *positive* 0-dynamically convex the proof works verbatim using positive equivariant symplectic homology.

Examples of contact manifolds satisfying the hypotheses of the previous theorem are given by the standard contact sphere S^{2n+1} and the unit cosphere bundle S^*S^{n+1} of S^{n+1}, with $n \geq 1$, endowed with the standard contact structure induced by the Liouville form on T^*S^{n+1}. Indeed, S^{2n+1} is a prequantization of $\mathbb{C}P^n$ and S^*S^{n+1} is a prequantization of $G_2^+(\mathbb{R}^{n+2})$, where $G_2^+(\mathbb{R}^{n+2})$ is the Grassmannian of oriented two-planes in $\mathbb{R}^{n+2}$ (see [8]). A computation shows that $c_{\mathbb{C}P^n} = n+1$, $c_{G_2^+(\mathbb{R}^{n+2})} \geq n$ (more precisely, $c_{G_2^+(\mathbb{R}^3)} = 2$ and $c_{G_2^+(\mathbb{R}^{n+2})} = n$ for $n > 1$), $\dim H_*(\mathbb{C}P^n; \mathbb{Q}) = n+1$, $\dim H_*(G_2^+(\mathbb{R}^{n+2}); \mathbb{Q}) = 2(\lfloor \frac{n}{2} \rfloor + 1)$ and $H_k(G_2^+(\mathbb{R}^{n+2}); \mathbb{Q}) = H_k(\mathbb{C}P^n; \mathbb{Q}) = 0$ for every odd k. Thus, we obtain the following immediate corollaries. The first one is for the standard contact sphere.

Corollary 16 ([9]) *Let α be a dynamically convex contact form on the standard contact sphere S^{2n+1} with $n \geq 1$. Then the Reeb flow of α carries at least two simple closed orbits. Moreover, if α is non-degenerate then it carries at least $n+1$ simple periodic orbits.*

The first statement generalizes the aforementioned result due to Ekeland and Hofer which establishes that every convex contact form on S^{2n+1} $(n > 0)$ has at least two geometrically distinct closed characteristics. The second statement was proved by Gutt and Kang [53] and generalizes the aforementioned result due to Long and Zhu [65].

The previous corollary was improved in the degenerate case by Ginzburg and Gurel [45] and Duan and Liu [33] independently. It generalizes the already mentioned results due to Long and Zhu [65] and Wang [87] to the dynamically convex scenario.

Theorem 17 ([33, 45]) *Let α be a dynamically convex contact form on the standard contact sphere S^{2n+1} with $n \geq 1$. Then the Reeb flow of α carries at least $\lceil (n+1)/2 \rceil + 1$ simple closed orbits. Moreover, if α is non-degenerate then it carries at least $n+1$ simple periodic orbits.*[1]

The second corollary, deals with the example of the unit cosphere bundle of spheres.

Corollary 18 ([9]) *Every 0-dynamically convex contact form α on S^*S^{n+1} endowed with the standard contact structure has at least two simple closed orbits. Moroever if α is non-degenerate then it has at least $2(\lfloor \frac{n}{2} \rfloor + 1)$ simple periodic orbits.*

Remark 19 The lower bounds of Corollaries 16 and 18 in the non-degenerate case are sharp as the irrational ellipsoid and the examples of Katok-Ziller show [90].

Remark 20 One can check (see [8, 9]) that in the case of S^*S^{n+1} the condition that α is 0-dynamically convex means that $\mu(\gamma) \geq k_-^0$ for every contractible closed orbit γ, where $k_-^0 = 3$ when $n = 1$ and $k_-^0 = n$ if $n > 1$.

As an application, consider a Finsler metric F on S^n. Denote by K the flag curvature of F and let $r := \max\{F(-v) \mid v \in TM,\ F(v) = 1\}$ be its reversibility. Suppose that F satisfies the pinching condition $(\frac{r}{r+1})^2 < K \leq 1$. We refer to [76] for the precise definitions; we just want to mention here that if F is a Riemannian metric then $r = 1$, K is the Gaussian curvature and the pinching condition reads as the classical assumption $1/4 < K \leq 1$. As observed in [8], the pinching condition implies that the contact form corresponding to F (whose Reeb flow is the geodesic flow) is 0-dynamically convex (see [54] in the case of S^2). Therefore, we obtain the following result.

Corollary 21 ([9]) *Let F be a Finsler metric on S^{n+1}, $n \geq 1$, satisfying the pinching condition $(\frac{r}{r+1})^2 < K \leq 1$. Then F has at least two prime closed geodesics. Moreover, if F is bumpy then it carries at least $2(\lfloor \frac{n}{2} \rfloor + 1)$ prime closed geodesics.*

The previous two corollaries can be improved in the degenerate case using the following result proved by Ginzburg and Gurel [45].

Theorem 22 ([45]) *Every 0-dynamically convex contact form α on S^*S^{n+1} endowed with the standard contact structure has at least $\lfloor \frac{n+1}{2} \rfloor - 1$ simple closed orbits. Moroever if α is non-degenerate then it has at least $2(\lfloor \frac{n}{2} \rfloor + 1)$ simple periodic orbits.*

The hypothesis of dynamical convexity is restrictive, but crucial in the proofs of the previous results. In [47], we relaxed considerably this assumption and the hypothesis on the minimal Chern number in the non-degenerate case. More precisely, we have

[1] Very recently, after this paper was accepted for publication, Çineli, Ginzburg, and Gürel [27] improved this lower bound in the degenerate case, showing that if α is dynamically convex then it carries at least $n+1$ simple periodic orbits, without assuming that α is non-degenerate.

the following result. We say that a periodic orbit γ is *good* if $\mu(\gamma) - \mu(\bar{\gamma}) \in 2\mathbb{Z}$ where $\bar{\gamma}$ is the underlying simple closed orbit. A contact form is *index-definite* if either every closed orbit has positive mean index or every closed orbit has negative mean index.

Theorem 23 ([47]) *Let (M^{2n+1}, ξ) be a prequantization S^1-bundle of a closed symplectic manifold (B, ω) such that $\omega|_{\pi_2(B)} \neq 0$, $c_B > n/2$ and $H_k(B; \mathbb{Q}) = 0$ for every odd k. Suppose that M admits a strong symplectic filling (W, Ω) such that $\Omega|_{\pi_2(W)} = 0$ and $c_1(TW)|_{\pi_2(W)} = 0$. Let α be a non-degenerate index-definite contact form supporting ξ. Assume that there is no good periodic orbit γ of α contractible in W such that*

- $\mu(\gamma) = 0$ *if n is odd;*
- $\mu(\gamma) \in \{0, \pm 1\}$ *if n is even.*

Then α carries at least r_B geometrically distinct periodic orbits contractible in W, where r_B is the total rank of $H_*(B; \mathbb{Q})$.

Remark 24 The proof is based on equivariant symplectic homology. The original statement in [47] is actually more general, although somewhat technical and involved, and includes the case that M is possibly not fillable.

Examples satisfying the hypotheses of the previous theorem include the standard contact sphere S^{2n+1} and the unit cosphere bundle of a compact rank one symmetric space (CROSS). More precisely, as already mentioned, S^{2n+1} is a prequantization of $\mathbb{C}P^n$, and its obvious filling in $\mathbb{R}^{2n+2}$ clearly satisfies the hypotheses above. A connected Riemannian manifold N is a symmetric space if for every $p \in N$ there exists an isometry $f_p : N \to N$ such that $f_p(p) = p$ and $f_p \circ \exp_p(v) = \exp_p(-v)$ for every $v \in T_pN$. The rank of a symmetric space N is the maximal dimension of a flat totally geodesic submanifold in N. By the classification of symmetric spaces, a CROSS is given by one of the following manifolds: S^m, $\mathbb{R}P^m$, $\mathbb{C}P^m$, $\mathbb{H}P^m$ and CaP^2; see [18] for details. Thus the filling of the unit cosphere bundle S^*N given by the unit codisk bundle D^*N in T^*N meets the condition of the previous theorem [47]. (Note that the examples are more general than those related to Theorem 14 due to the weakening of the hypothesis on the minimal Chern number c_B.)

Every CROSS N admits a metric such that all of its geodesics are periodic with the same minimal period; in other words, the geodesic flow generates a free circle action on S^*N. Thus the unit cosphere bundle S^*N is a prequantization of a closed symplectic manifold (B, ω). Moreover, a homological computation shows that $H_k(B; \mathbb{Q}) = 0$ for every odd k; see [90, page 141]. In this case, the total rank r_B of $H_*(B; \mathbb{Q})$ and the minimal Chern number c_B are given by Table 1.

Notice that the hypothesis on c_B in the previous theorem barely holds for $M = S^*\mathbb{C}P^m$, where $\dim B/4 = m - 1/2$ and $c_B = m$. We have the following corollary, which was previously proved for the standard contact sphere by Duan, Liu, Long and Wang in [34] and for Finsler metrics on a simply connected CROSS by Duan, Long and Wang in [36].

Table 1 Prequantizations and the corresponding r_B and c_B

Prequantization	$r_B = \dim H_*(B; \mathbb{Q})$	c_B
S^{2n+1}	$n+1$	$n+1$
S^*S^2 or $S^*\mathbb{R}P^2$	2	2
S^*S^m or $S^*\mathbb{R}P^m$ with $m > 2$ even	m	$m-1$
S^*S^m or $S^*\mathbb{R}P^m$ with m odd	$m+1$	$m-1$
$S^*\mathbb{C}P^m$	$m(m+1)$	m
$S^*\mathbb{H}P^m$	$2m(m+1)$	$2m+1$
S^*CaP^2	24	11

Corollary 25 ([47]) *Let (M, ξ) be either the standard contact sphere S^{2n+1} or the unit cosphere bundle S^*N of a CROSS and let α be a contact form supporting ξ. Assume that α satisfies the condition of the previous theorem. Then α has at least r_B geometrically distinct periodic orbits, where r_B is given by Table 1.*

The prequantization bundles in Corollary 25 admit contact forms with precisely r_B geometrically distinct periodic orbits. These contact forms are given by irrational ellipsoids and the Katok-Ziller Finsler metrics; [90] (cf. Remark 19). This shows that the lower bound in Theorem 23 is sharp.

2.2.2 Multiplicity Results Assuming Convexity and Symmetry

As mentioned before, a long standing conjecture in Hamiltonian Dynamics states that every contact form α on S^{2n+1} has at least $n+1$ simple closed Reeb orbits and a remarkable result due to Long and Zhu [65], and improved by Wang [87], establishes that every *convex* contact form α on S^{2n+1} has at least $\lceil (n+1)/2 \rceil + 1$ simple closed Reeb orbits. This result was extended to *dynamically convex* contact forms by Ginzburg-Gurel [45] and Duan-Liu [33] independently (see Theorem 17).

When α is convex *and invariant by the antipodal map*, it was proved by Liu, Long and Zhu [60] that there are $n+1$ simple closed Reeb orbits, showing that the aforementioned conjecture is true for this important class of contact forms. The generalization of this result replacing the hypothesis of convexity by a symplectic condition was done in [49] and involves a subtle modification of the notion of dynamical convexity, called *strong dynamical convexity*, which takes into account the action of the symmetry group.

Before we give the precise definition, we need to introduce some terminology. Let $A \in \mathrm{Sp}(2d)$ be a totally degenerate symplectic matrix, i.e., all eigenvalues of A are equal to one. One can show that then $A = \exp(JQ)$, where J is the canonical complex structure and Q is a symmetric matrix with all eigenvalues zero. Examining Williamson's normal forms (see, e.g., [14, Appendix 6]) it is easy to see that the quadratic form Q can be symplectically decomposed into a sum of terms of four types:

- the identically zero quadratic form on $\mathbb{R}^{2d}$,
- the quadratic form $Q_0 = p_1q_2 + p_2q_3 + \cdots + p_{d-1}q_d$ in Darboux coordinates on $\mathbb{R}^{2d}$, where $d \geq 3$ is odd,
- the quadratic forms $Q_\pm = \pm(Q_0 + p_d^2/2)$ on $\mathbb{R}^{2d}$ for any d, where $Q_0 = 0$ when $d = 1$.

We define $b_\pm(A)$ as the number of $Q_\pm$ terms in this decomposition. These are linear symplectic invariants of A although this is not obvious; cf. [45]. Furthermore, given a symplectic matrix P, let V be the subspace whose complexification is the generalized eigenspace of the eigenvalue one. Define $b_\pm(P) = b_\pm(P|_V)$. For a closed Reeb orbit γ, we set $b_\pm(\gamma) = b_\pm(P_\gamma)$, where P_γ is the linearized Poincaré return map of γ.

Let G be a group acting on S^{2n+1} and let α be a G-invariant contact form. A closed Reeb orbit γ of α is called symmetric if $g(\gamma(\mathbb{R})) = \gamma(\mathbb{R})$ for every $g \in G$. Given a G-invariant contact form, let $\mathcal{P}$ be the set of simple closed Reeb orbits and write $\mathcal{P} = \mathcal{P}_s \cup \mathcal{P}_{ns}$ where $\mathcal{P}_s$ (resp. $\mathcal{P}_{ns}$) is the set of symmetric (resp. non-symmetric) simple closed orbits.

Definition 26 ([49]) A contact form α is *strongly dynamically convex* if

$$\mu(\gamma) \geq n + 2$$

for every $\gamma \in \mathcal{P}$, and

$$\mu(\gamma_s) + b_-(\gamma_s) - b_+(\gamma_s) \geq n + 2$$

for every $\gamma_s \in \mathcal{P}_s$.

Note that this notion is natural from the point of view of contact topology since it is invariant under contactomorphisms in the following sense. Let $\varphi : S^{2n+1} \to S^{2n+1}$ be a contactomorphism and consider a G-action on S^{2n+1}. If α is a G-invariant strongly dynamically convex contact form then $\varphi^*\alpha$ is strongly dynamically convex under the conjugated G-action $\varphi^{-1}g\varphi$.

The first result in [49] establishes that every G-invariant contact form on S^{2n+1} satisfying a condition slightly weaker than strong dynamical convexity carries at least $n + 1$ simple periodic orbits. (Note the difference between the lower bound $n + 1$ in the theorem below and the lower bound $n + 2$ in Definition 26.) More precisely, let $\mathcal{P}^+ \subset \mathcal{P}$ be the set of simple periodic orbits with positive mean index. Write $\mathcal{P}^+ = \mathcal{P}^+_s \cup \mathcal{P}^+_{ns}$ where $\mathcal{P}^+_s$ (resp. $\mathcal{P}^+_{ns}$) is the set of symmetric (resp. non-symmetric) closed orbits with positive mean index.

Theorem 27 ([49]) *Let G be a group acting on S^{2n+1} and α a G-invariant contact form. Suppose that α is strongly dynamically convex or more generally*

$$\mu(\gamma) \geq n + 1$$

for every $\gamma \in \mathcal{P}^+$ *and*

$$\mu(\gamma_s) + b_-(\gamma_s) - b_+(\gamma_s) \geq n + 1$$

for every $\gamma_s \in \mathcal{P}_s^+$. *Then* $\#\mathcal{P}^+ \geq n + 1$.

Remark 28 In fact, the result is more general than stated. We do not really need the action of a group: it is enough to consider a *subset* of $Diff(S^{2n+1})$ that is not necessarily a subgroup. More precisely, given a subset $S \subset Diff(S^{2n+1})$ we say that α is S-invariant if $\varphi^*\alpha = \alpha$ for every $\varphi \in S$. A closed orbit γ of α is symmetric if $\varphi(\gamma(\mathbb{R})) = \gamma(\mathbb{R})$ for every $\varphi \in S$. Then the previous theorem holds for S-invariant contact forms. Moreover, the definition of strong dynamical convexity clearly extends to this more general context.

In particular, when the action is trivial, we have the following consequence of the theorem.

Corollary 29 ([49]) *Let* α *be a contact form on* S^{2n+1}. *Suppose that*

$$\mu(\gamma) \geq n + 1 \quad \textit{and} \quad \mu(\gamma) + b_-(\gamma) - b_+(\gamma) \geq n + 1$$

for every $\gamma \in \mathcal{P}^+$. *Then* $\#\mathcal{P}^+ \geq n + 1$.

Clearly, $b_\pm(P) = 0$ for every symplectic matrix P such that the eigenvalue one is semisimple. Therefore, we obtain the following corollary.

Corollary 30 ([49]) *Let* α *be a dynamically convex contact form on* S^{2n+1}. *Suppose that the eigenvalue one of the linearized Poincaré map of every degenerate simple closed orbit of* α *is semisimple. Then* $\#\mathcal{P} \geq n + 1$. *In particular, this lower bound holds if every simple closed orbit of* α *is non-degenerate.*

Remark 31 Note that the condition that every *simple* periodic orbit is non-degenerate (but iterated orbits may degenerate) is considerably weaker than the assumption that α is non-degenerate.

Unfortunately, when the action is trivial, it is probably not true that every convex contact form is strongly dynamically convex. However, for some natural non-trivial actions, convex invariant contact forms are strongly dynamically convex. More precisely, we have the following result.

Theorem 32 ([49]) *Let* α *be a convex contact form on* S^{2n+1} *invariant under the antipodal map. Then* α *is strongly dynamically convex.*

As a consequence, we recover the aforementioned result from [60]:

Corollary 33 *Every convex contact form on* S^{2n+1} *invariant under the antipodal map has at least* $n + 1$ *simple closed orbits.*

2.2.3 Multiplicity Results Assuming only Non-degeneracy

Let M^{2n+1} be a closed contact manifold admitting a strong symplectic filling W such that $c_1(TW) = 0$. In [5], we establish that a contact form α on M has at least *two* simple closed orbits provided that the positive equivariant symplectic homology $\mathrm{HC}_*(W)$ satisfies a mild condition and assuming *only* the non-degeneracy of α. More precisely, we have the following statement. In what follows, we will use the convention that the natural numbers are given by the positive integers. Moreover, the grading of $\mathrm{HC}_*(W)$ is given by the Conley-Zehnder index. Under the assumption that $c_1(TW) = 0$, this grading is well defined up to a choice of the homotopy class of a non-vanishing section of the determinant line bundle $\Lambda_{\mathbb{C}}^{n+1}TW$; see [10, Sect. 3], [67, 80]. For a shorter notation, we will omit this dependence.

Theorem 34 ([5]) *Let (M^{2n+1}, ξ) be a closed contact manifold admitting a strong symplectic filling W such that $c_1(TW) = 0$. Let Γ be a set of free homotopy classes of loops in W closed under iterations and assume that there exist $K \in \mathbb{N}$ and a non-vanishing section σ of the determinant line bundle $\Lambda_{\mathbb{C}}^{n+1}TW$ such that*

$$\dim \mathrm{HC}_n^\Gamma(W) < \dim \mathrm{HC}_{n+jK}^\Gamma(W)$$

or

$$\dim \mathrm{HC}_{-n}^\Gamma(W) < \dim \mathrm{HC}_{-n-jK}^\Gamma(W)$$

for every $j \in \mathbb{N}$, where the grading in $\mathrm{HC}_*^\Gamma(W)$ is taken with respect to the homotopy class of σ. Then every non-degenerate Reeb flow on M carries either infinitely many geometrically distinct closed Reeb orbits or at least two geometrically distinct closed Reeb orbits γ_1 and γ_2 such that their Conley-Zehnder indices satisfy $\mu(\gamma_1^k) \neq \mu(\gamma_2^k)$ for some $k \in \mathbb{N}$. Moreover, all these orbits have free homotopy class in Γ.

Remark 35 When $\Gamma = \{0\}$, i.e. when looking at contractible orbits, it is enough to assume that $c_1(TW)|_{\pi_2(W)} = 0$. Moreover, in this case the grading of $\mathrm{HC}_*^\Gamma(W)$ does not depend on the choice of the homotopy class of σ.

Let $M \xrightarrow{\tau} \overline{M}$ be a finite covering. Suppose that M carries a contact structure ξ preserved by the deck transformations so that $\overline{M}$ has an induced contact structure $\overline{\xi}$ such that $\tau^*\overline{\xi} = \xi$. We say in this case that $\overline{M}$ is a *contact finite quotient* of M. Assume that (M, ξ) satisfies the hypotheses of Theorem 34 and that $H^1(M; \mathbb{R}) = 0$. Given a non-degenerate contact form $\overline{\alpha}$ on $\overline{M}$ we can consider its lift to M and applying Theorem 34 we can deduce that $\overline{\alpha}$ has at least two simple closed orbits; see [5, Sect. 4]. Thus, we have the following corollary.

Corollary 36 ([5]) *Let (M, ξ) be a contact manifold satisfying the hypotheses of Theorem 34 and $H^1(M; \mathbb{R}) = 0$. Then every non-degenerate Reeb flow on a contact finite quotient of M carries at least two simple closed orbits. Moreover, the closed lifts of iterates of these orbits to M have free homotopy class in W contained in Γ.*

Remark 37 The hypothesis that $H^1(M;\mathbb{R}) = 0$ can be dropped if the first Chern class of the contact structure on the quotient vanishes.

Remark 38 When $\Gamma = \{0\}$, the hypothesis that $H^1(M;\mathbb{R}) = 0$ can be replaced by the assumption that the map $\pi_1(M) \to \pi_1(W)$ induced by the inclusion is injective.

Remark 39 Since the deck transformations do not necessarily extend to the filling W, it is not clear that when $\Gamma = \{0\}$ the assumption that $c_1(TW)|_{\pi_2(W)} = 0$ is enough to conclude the previous corollary, unless the map $\pi_1(M) \to \pi_1(W)$ induced by the inclusion is injective; cf. Remark 35.

In [5] we provided several examples of contact manifolds satisfying the hypotheses of Theorem 34. As a matter of fact, we have the following consequences of Theorem 34 whose proofs are given in [5]. For a definition of good toric contact manifolds, see, for instance, [7, 59].

Corollary 40 (Displaceable contact manifolds [5]) *Let (M, ξ) be a contact manifold admitting a displaceable exact contact embedding into a convex at infinity exact symplectic manifold X such that $c_1(TX)|_{\pi_2(X)} = 0$ and denote by W the compact region in X bounded by M. Then every non-degenerate Reeb flow on M has at least two simple closed orbits contractible in W. If $c_1(TX) = 0$ and $H^1(M;\mathbb{R}) = 0$ then every Reeb flow on a contact finite quotient of M carries at least two simple closed orbits. Moreover, the closed lifts of iterates of these orbits to M are contractible in W.*

Corollary 41 (Unit cosphere bundles [5]) *Let N be a closed oriented spin manifold with dimension bigger than one such that $\pi_1(N)$ is finite. Then every non-degenerate contact form on (S^*N, ξ_{can}) has at least two simple closed orbits.*

Corollary 42 (Good toric contact manifolds [5]) *Let (M, ξ) be a good toric contact manifold admitting a strong symplectic filling W such that $c_1(TW) = 0$. Then every non-degenerate contact form on a contact finite quotient of M carries at least two geometrically distinct contractible closed orbits.*

Corollary 43 (Prequantization circle bundles [5]) *Let (M, ξ) be a prequantization circle bundle of a closed integral symplectic manifold (B, ω) such that $\omega|_{\pi_2(B)} \neq 0$, $c_1(TB)|_{\pi_2(B)} \neq 0$ and, furthermore, $H_k(B;\mathbb{Q}) = 0$ for every odd k or $c_B > n$. Suppose that M admits a strong symplectic filling W such that $c_1(TW) = 0$. Then every non-degenerate Reeb flow on M carries at least two geometrically distinct closed orbits contractible in W. If, additionally, $H^1(M;\mathbb{R}) = 0$ then every contact form on a contact finite quotient of M carries at least two geometrically distinct closed orbits. Moreover, the closed lifts of iterates of these orbits to M are contractible in W.*

Corollary 44 (Brieskorn spheres [5]) *Let M be a contact finite quotient of a Brieskorn sphere. Then every non-degenerate Reeb flow on M carries at least two geometrically distinct closed orbits.*

Moreover, connected sums of contact manifolds meeting the assumptions of Theorem 34 also satisfy these assumptions as the following result shows.

Theorem 45 (Connected sums [5]) *Let* (W_1, λ_1) *and* (W_2, λ_2) *be Liouville domains of dimension* $2n+2$ *with vanishing first Chern class. Assume that there are non-vanishing sections* σ_1 *and* σ_2 *of* $\Lambda_{\mathbb{C}}^{n+1} TW_1$ *and* $\Lambda_{\mathbb{C}}^{n+1} TW_2$ *respectively satisfying the hypothesis of Theorem 34 with* Γ *given by the set of all free homotopy classes. Suppose that* $c_1(T(W_1 \# W_2)) = 0$ *and let* σ *be a non-vanishing section of* $\Lambda_{\mathbb{C}}^{n+1} T(W_1 \# W_2)$ *extending* σ_1 *and* σ_2*. Then* $W_1 \# W_2$ *satisfies the hypothesis of Theorem 34 with the grading of* $\mathrm{HC}_*(W_1 \# W_2)$ *induced by* σ*.*

2.2.4 Periodic Orbits of Non-degenerate Lacunary Contact Forms

As noticed at the end of Sect. 2.2.1, the prequantization bundles in Corollary 25 admit contact forms with precisely $r_B = \dim H_*(B; \mathbb{Q})$ geometrically distinct periodic orbits. These contact forms are given by irrational ellipsoids and the Katok-Ziller Finsler metrics [90]. There are several other examples of contact manifolds admitting contact forms with finitely many closed orbits, like, for instance, good toric contact manifolds [7] and prequantizations of closed integral symplectic manifolds admitting Hamiltonian circle actions with isolated fixed points (one can lift the Hamiltonian circle action on the basis to a Reeb flow on the prequantization such that the closed orbits correspond precisely to the fibers over the fixed points).

To the best of the author's knowledge, every contact form with finitely many simple closed orbits known so far is non-degenerate. Thus, Theorem 23 raises the following very hard question:

Question
Let (M, ξ) be a prequantization S^1-bundle of a closed integral symplectic manifold B. Is it true that every contact form α supporting ξ carries at least r_B simple closed orbits?

For instance, in the particular case of the standard contact sphere S^{2n+1} this question is the aforementioned long standing conjecture that every contact form has at least $n+1$ simple closed orbits which is true when $n=1$ but far from being known when $n>1$.

To the best of the author's knowledge, every contact form with finitely many simple closed orbits on a prequantization S^1-bundle known so far has *precisely* r_B *closed orbits*. Thus, we can ask the even harder question:

Question
Let (M, ξ) be a prequantization S^1-bundle of a closed integral symplectic manifold B. Is it true that every contact form α supporting ξ has either r_B or infinitely many simple closed orbits?

It was recently proved that it is true in the particular case that $M = S^3$ by Cristofaro-Gardiner-Hutchings-Hryniewicz-Liu [28]. It was proved before for dynamically convex contact forms (degenerate or not) on S^3 by Hofer-Wysocki-Zehnder [56] and for non-degenerate contact forms on S^3 by Cristofaro-Gardiner-Hutchings-Pomerleano [30].

We say that a contact form is *lacunary* if the indexes of every contractible closed orbit have the same parity. To the best of the author's knowledge, every contact form with finitely many simple closed orbits known so far is lacunary. As a matter of fact, in all the examples every periodic orbit is elliptic and non-degenerate and it turns out that the index of an elliptic non-degenerate periodic orbit on a contact manifold M^{2n+1} has the same parity of n.

Motivated by these considerations, we proved the following result. In what follows, we say that a contractible closed orbit is simple if it is not an iterate of another contractible orbit, although it maybe the iterate of a non-contractible one.

Theorem 46 ([12]) *Let (M^{2n+1}, ξ) be a prequantization S^1-bundle of a closed symplectic manifold (B, ω) such that $\omega|_{\pi_2(B)} \neq 0$, $c_B > n/2$ and $H_k(B; \mathbb{Q}) = 0$ for every odd k. Assume also that M is spherically positive monotone, that is, $c_1(TB)|_{\pi_2(B)} = \lambda\omega|_{\pi_2(B)}$ for some $\lambda > 0$. Suppose that M admits a strong symplectic filling W such that $c_1(TW)|_{\pi_2(W)} = 0$ and the map $\pi_1(M) \to \pi_1(W)$ induced by the inclusion is injective. Let α be a non-degenerate lacunary contact form supporting ξ. Then α has precisely r_B simple contractible closed orbits, where $r_B = \dim H_*(B; \mathbb{Q})$.*

Remark 47 The statement in [12] is actually more general and includes the case that M is possibly not fillable. Moreover, in [12] it is also considered some prequantizations of orbifolds.

Now, note that when $\pi_1(M)$ is finite, the simple contractible closed orbits are in bijection with the simple closed orbits. Therefore, we obtain the following corollary.

Corollary 48 ([12]) *Let (M^{2n+1}, ξ) be a prequantization S^1-bundle of a closed symplectic manifold (B, ω) and α a non-degenerate lacunary contact form supporting ξ as in Theorem 46. Assume that $\pi_1(M)$ is finite. Then α has precisely r_B simple closed orbits, where $r_B = \dim H_*(B; \mathbb{Q})$.*

Now, recall from Sect. 2.2.1 that the standard contact sphere and the unit cosphere bundle of a CROSS (of dimension bigger than two) satisfy the hypotheses of Theorem 46. (It is easy to see that these examples are prequantizations of spherically positive monotone symplectic manifolds.) Moreover, their fundamental groups are clearly finite. Therefore, we have the following result (cf. Corollary 25).

Corollary 49 ([12]) *Let (M^{2n+1}, ξ) be either the standard contact sphere S^{2n+1} or the unit cosphere bundle S^*N of a CROSS and let α be a non-degenerate lacunary contact form supporting ξ. Then α has precisely r_B simple closed orbits, where r_B is given by Table 1.*

The previous corollary improves results due to Duan, Liu and Ren [35] for the standard contact sphere and Duan and Xie [37] for Finsler metrics on a simply connected CROSS.

3 Stability of Periodic Orbits for Reeb Flows

A closed Reeb orbit is called *elliptic*[2] or *stable* if every eigenvalue of its linearized Poincaré map has modulus one. The existence of elliptic orbits has important consequences for the Reeb flow. As a matter of fact, fundamental contributions due to Arnold, Birkhoff, Kolmogorov, Moser, Newhouse, Smale and Zehnder establish that, under generic conditions, the presence of elliptic orbits implies a rich dynamics: it forces the existence of a subset of positive measure filled by invariant tori (this implies, in particular, that the flow is not ergodic with respect to the volume measure), existence of transversal homoclinic connections and positivity of the topological entropy; see [4] and references therein.

The search for elliptic closed orbits goes back to Poincaré [71] and a long-standing conjecture in Hamiltonian Dynamics states, in modern terminology, that the Reeb flow of every contact form on the standard contact sphere carries an elliptic periodic orbit [39, 64]. By the dynamical consequences of the existence of elliptic orbits mentioned before, this conjecture leads to deep implications: if it holds then Boltzmann's ergodic hypothesis would fail for a large class of important and natural Hamiltonian systems.

Unfortunately this is still an open question, but there are some important partial results. The first one, proved by Ekeland [38, 39], asserts the existence of an elliptic closed orbit on a regular energy level of a convex Hamiltonian in $\mathbb{R}^{2n}$ satisfying suitable *pinching conditions* on the Hessian. The second one, due to Dell'Antonio-D'Onofrio-Ekeland [32], establishes that if a hypersurface in $\mathbb{R}^{2n}$ is convex (i.e. diffeomorphic to a sphere and bounding a strictly convex subset) and *invariant by the antipodal map* then its Reeb flow carries an elliptic closed orbit. The proofs are based on classical variational methods and use in a strong way the hypothesis of convexity.

In [8], these results are generalized for contact manifolds using Symplectic Field Theory and a neck-stretching argument. The first main result establishes that, under some hypotheses, dynamically convex contact forms have elliptic orbits with extremal Conley-Zehnder indexes. In order to state it, we need some preliminary definitions. We say that a Morse function is *even* if every critical point has even

[2] Some authors rather call this an *elliptic-parabolic* periodic orbit.

Morse index. Given a prequantization S^1-bundle M, denote by β the contact form (given by the connection form) whose Reeb flow generates the corresponding free circle action. Given a finite subgroup $G \subset S^1$, we say that a contact form α on M is *G-invariant* if $(\phi_\beta^{t_0})^*\alpha = \alpha$, where ϕ_β^t is the flow of R_β and $t_0 \in S^1$ is a generator of G. A periodic orbit γ of a G-invariant contact form α is called *symmetric* if $\phi_\beta^{t_0}(\gamma(\mathbb{R})) = \gamma(\mathbb{R})$. In what follows, given a closed orbit γ, we denote by $\mu_{RS}(\gamma)$ its Robbin-Salamon index; see [78].

Theorem 50 ([8]) *Let (M^{2n+1}, ξ) be a prequantization S^1-bundle of a closed symplectic manifold (B, ω) such that $c_1(TB) = \lambda\omega$ for some constant $\lambda \in \mathbb{R}$. Let φ be a simple orbit of R_β and denote by a the free homotopy class of φ. Assume that one of the following two conditions holds:*

(H1) *B admits an even Morse function, $a = 0$ and $\lambda > 1$.*
(H2) *$\omega|_{\pi_2(B)} = 0$ and $\pi_1(B)$ is torsion free.*

Let G be a finite subgroup of S^1 and α a G-invariant contact form supporting ξ. Then α carries a symmetric closed orbit $\tilde{\gamma}$ with free homotopy class a. Moreover, if G is non-trivial and every periodic orbit γ of α with free homotopy class a satisfies $\mu(\gamma) \geq \mu_{RS}(\varphi) - n$ in case (H1) or $\mu^+(\gamma) \leq \mu_{RS}(\varphi) + n$ in case (H2) then $\tilde{\gamma}$ is elliptic and its Conley-Zehnder index satisfies $\mu(\tilde{\gamma}) = \mu_{RS}(\varphi) - n$ in case (H1) or $\mu^+(\tilde{\gamma}) = \mu_{RS}(\varphi) + n$ in case (H2).

The statement above can be nicely rephrased using dynamical convexity, as in Definition 10, in the following way. Under the assumptions of Theorem 50, one can construct a non-degenerate contact form β' (given by a perturbation of β) supporting ξ for which cylindrical contact homology is well defined (that is, its differential ∂ is well defined and satisfies $\partial^2 = 0$) and $k_-^a = \mu_{RS}(\varphi) - n$ in case (H1) and $k_+^a = \mu_{RS}(\varphi) + n$ in case (H2), with k_-^a and k_+^a as in Definition 10. Thus, if G is non-trivial and α is positively a-dynamically convex in case (H1) or negatively a-dynamically convex in case (H2) then α carries an elliptic symmetric closed orbit. Theorem 50 is not stated in this way because contact homology is yet to be fully put on a rigorous basis.

Remark 51 It can be shown that $\tilde{\gamma}$ is actually strictly elliptic, that is, $\tilde{\gamma}$ is elliptic and every eigenvalue of its linearized Poincaré map different from one is Krein-definite.

Remark 52 The hypothesis that $c_1(TB) = \lambda\omega$ for some constant $\lambda \in \mathbb{R}$ is equivalent to the condition that $c_1(\xi)|_{H_2(M,\mathbb{Z})} = 0$. The hypothesis that $\pi_1(B)$ is torsion free is used only to ensure that, under the hypothesis (H2), the free homotopy class of the fiber is primitive. If $\pi_r(B) = 0$ for every $r \geq 2$ then (H2) is automatically satisfied; see [46].

Remark 53 The indexes of γ and φ are computed using a trivialization of ξ over a closed curve representing the free homotopy class a (see [8, Sect. 3]). It is easy to see that the assumptions on α do not depend on the choice of this trivialization: the differences $\mu_{RS}(\varphi) - \mu(\tilde{\gamma})$ and $\mu_{RS}(\varphi) - \mu^+(\tilde{\gamma})$ do not depend on this, see [8, Equation (3.10)].

Remark 54 Theorem 50 implies, in particular, that the Weinstein conjecture holds for contact manifolds satisfying the hypotheses (H1) or (H2) using techniques of contact homology without transversality issues. Notice that M does not need to admit an aspherical symplectic filling. In fact, in the proof we do not need a filling of M at all. This is important because in the proof we consider the contact homology of the quotient M/G which, in general, does not admit an aspherical symplectic filling.

Remark 55 In (H1), the hypotheses on λ can be replaced by the weaker assumption that $\lambda > 0$ if we assume that the relevant moduli spaces of pseudo-holomorphic curves used to define and prove the invariance of the contact homology can be cut out transversally. (This condition is used only to assure that in case (H1) a positively a-dynamically convex contact form has the property that every contractible periodic orbit has non-negative reduced Conley-Zehnder index.) Moreover, the argument is easier in this case, see [8]. However, with the hypotheses of Theorem 50, its proof is rigorously established, with no transversality issues. A crucial ingredient for this is the fact that the orbit used in the neck-stretching argument developed in [8, Sect. 5.2.2] has minimal index and action in case (H1) and maximal index and action in case (H2). This ensures that we have to deal only with somewhere injective holomorphic curves.

As noticed before, S^{2n+1} with the standard contact structure is a prequantization of $\mathbb{C}P^n$ and its first Chern class vanishes. Moreover, $\mathbb{C}P^n$ clearly admits an even Morse function (as any toric symplectic manifold) and the Robbin-Salamon index of the (simple) orbits of the circle action is equal to $2n+2$ (for a trivialization given by a capping disk). The action of $\mathbb{Z}_2 \subset S^1$ is generated by the antipodal map. Consequently we obtain the following generalization of the result proved by Dell'Antonio, D'Onofrio and Ekeland [32] mentioned before.

Corollary 56 ([8]) *If a contact form on S^{2n+1} is dynamically convex (that is, every periodic orbit γ satisfies $\mu(\gamma) \geq n+2$) and invariant by the antipodal map then its Reeb flow carries an elliptic symmetric closed orbit γ. Moreover, $\mu(\gamma) = n+2$.*

Remark 57 Clearly the $\mathbb{Z}_2$-invariance of the contact form in the previous corollary can be generalized to $\mathbb{Z}_m$-invariance for any $m \geq 2$. This means that if a starshaped hypersurface S in $\mathbb{R}^{2n+2}$ is dynamically convex and invariant by the map $z \mapsto e^{2\pi i/m}z$ then it carries an elliptic symmetric closed characteristic. If S is convex then this result follows from a generalization of [32] due to Arnaud [13]; see also [11]. Moreover, Long and Zhu [65] proved the existence of an elliptic closed characteristic on a convex hypersurface S in $\mathbb{R}^{2n+2}$, without any hypothesis of invariance, under the restrictive assumption that its Reeb flow carries only finitely many simple periodic orbits. Under more restrictive assumptions, several results have been obtained concerning the multiplicity of the elliptic periodic orbits; see, for instance, [63, 64, 84, 85] and references therein.

The second main result in [8] shows that the hypothesis of non-triviality of G used to ensure that the periodic orbit given by Theorem 50 is elliptic can be relaxed once we impose certain bounds on the Conley-Zehnder indexes of some high iterate of the periodic orbits.

Theorem 58 ([8]) *Let* (M^{2n+1}, ξ) *be a contact manifold satisfying the hypotheses of Theorem 50 and* α *a contact form supporting* ξ. *Suppose that there exists an integer* $k > 1$ *such that every periodic orbit* γ *of* α *with free homotopy class a satisfies* $\mu(\gamma) \geq \mu_{RS}(\varphi) - n$ *and* $\mu(\gamma^k) \geq k\mu_{RS}(\varphi) - n$ *in case (H1) or* $\mu^+(\gamma) \leq \mu_{RS}(\varphi) + n$ *and* $\mu^+(\gamma^k) \leq k\mu_{RS}(\varphi) + n$ *in case (H2). Then* α *carries an elliptic periodic orbit* $\tilde{\gamma}$. *Moreover,* $\mu(\tilde{\gamma}) = \mu_{RS}(\varphi) - n$ *in case (H1) and* $\mu^+(\tilde{\gamma}) = \mu_{RS}(\varphi) + n$ *in case (H2).*

Remark 59 As in Remark 51, we actually have that $\tilde{\gamma}$ is strictly elliptic.

The hypothesis in Theorem 58 has the following subtle homological meaning. In dynamical convexity, the inequalities $\mu(\gamma) \geq k_-$ and $\mu(\gamma) \leq k_+$ are obvious necessary conditions for a non-degenerate closed orbit γ being homologically essential: clearly γ cannot contribute to the contact homology if one of these inequalities is violated due to the very definition of k_- and k_+. On the other hand, as explained in [8, Sect. 5.2.1], under the assumptions of Theorem 50, the contact homology of a prequantization M of a closed symplectic manifold (B, ω) is given by copies of the singular homology of B with a shift in the degree. More precisely, we have the isomorphism

$$\mathrm{HC}_*(M) \cong \oplus_{k\in\mathbb{N}} H_{*+n-k\mu_{RS}(\varphi)}(B; \mathbb{Q}),$$

see [8, Eq. (5.4)]. Thus, the inequalities $\mu(\gamma^k) \geq k\mu_{RS}(\varphi) - n$ and $\mu(\gamma^k) \leq k\mu_{RS}(\varphi) + n$ are obvious necessary conditions for γ^k being homologically essential *in the k-th copy of* $H_*(B; \mathbb{Q})$ *in the contact homology.*

Now, let M be an hypersurface in $\mathbb{R}^{2n+2}$ such that $M = H^{-1}(1/2)$, where $H : \mathbb{R}^{2n+2} \to \mathbb{R}$ is a convex Hamiltonian homogeneous of degree two. As defined in [38, 39], we say that M is (r, R)*-pinched*, with $0 < r \leq R$, if

$$\|v\|^2 R^{-2} \leq \langle d^2 H(x)v, v\rangle \leq \|v\|^2 r^{-2}$$

for every $x \in M$ and $v \in \mathbb{R}^{2n+2}$, where $d^2H(x)$ is the Hessian of H at x. Using results due to Croke and Weinstein [31] and lower estimates for the Conley-Zehnder index, we can show that if M is (r, R)-pinched with $\frac{R}{r} < \sqrt{\frac{k}{k-1}}$ for some real number $k > 1$ then every periodic orbit γ of M satisfies $\mu(\gamma^{\lfloor k \rfloor}) \geq \lfloor k \rfloor(2n+2) - n$, where $\lfloor k \rfloor = \max\{j \in \mathbb{Z} \mid j \leq k\}$ (note that when $k \to 1$ we conclude that every convex hypersurface is dynamically convex); see [8, Section 3.4]. Thus, we derive the following aforementioned result due to Ekeland [38, 39].

Corollary 60 *Let* M *be a* (r, R)*-pinched convex hypersurface in* $\mathbb{R}^{2n+2}$ *with* $R/r < \sqrt{2}$. *Then* M *carries an elliptic closed Reeb orbit* γ *such that* $\mu(\gamma) = n + 2$.

Remark 61 It can be shown that the elliptic orbit given by the previous corollary is simple.

3.1 Applications to Geodesic and Magnetic Flows

The hypothesis of non-triviality of G in Theorem 50 is probably just technical, but it is crucial in our argument. However, in many applications this condition is automatically achieved due to topological reasons. Let us start illustrating this with an application to geodesic flows on S^2. Let F be a Finsler metric on S^2 with reversibility r. It is well known that the geodesic flow of any Finsler metric on S^2 lifts to a Hamiltonian flow on a suitable star-shaped hypersurface in $\mathbb{R}^4$ via a double cover (the sphere bundle SS^2 is contactomorphic to $\mathbb{R}P^3$ with the contact structure induced by S^3 tight). As mentioned in Sect. 2.2.1, it is proved in [54] that, if the flag curvature K of F satisfies the pinching condition $(r/(r+1))^2 < K \leq 1$, then the lift of the geodesic flow to S^3 is dynamically convex. Thus, we have the following theorem. It was proved before by Rademacher [77] using different methods and assuming that the pinching condition is strict.

Theorem 62 ([8]) *Let F be a Finsler metric on S^2 with reversibility r and flag curvature K satisfying $(r/(r+1))^2 \leq K \leq 1$. Then F carries an elliptic closed geodesic γ whose Morse index satisfies $\mu_{Morse}(\gamma) \leq 1$. Moreover, if $(r/(r+1))^2 < K \leq 1$ then γ is prime and $\mu_{Morse}(\gamma) = 1$.*

The case where the inequality $(r/(r+1))^2 \leq K$ is not strict can be dealt with using a perturbation argument; see [8] for details.

Applications to geodesic flows in higher dimensions are given by the next two results. Consider a Finsler metric F on S^{n+1} with $n \geq 1$, reversibility r and flag curvature K. An easy computation of the contact homology of the unit sphere bundle SS^{n+1} shows that $k_- = \mu_{RS}(\varphi) - n = n$, where φ is a prime closed geodesic of the round metric and $\mu_{RS}(\varphi)$ is computed using a suitable trivialization. Results due to Rademacher [76] establish that if F satisfies the pinching condition $(r/(r+1))^2 < K \leq 1$ then every closed geodesic γ of F satisfies $\mu(\gamma) \geq n$. In other words, this pinching condition implies that the contact form defining the geodesic flow is positively dynamically convex. Moreover, SS^{n+1} is a prequantization of the Grassmannian of oriented two-planes $G_2^+(\mathbb{R}^{n+2})$ which admits an even Morse function and is monotone with constant of monotonicity $\lambda > 1$. Therefore, we get the following result which generalizes [16, Theorem A(ii)]. It was also proved by Rademacher [77] using different methods and assuming that the pinching condition is strict. In the statement, K_- and K_+ are continuous functions defined on the unit sphere bundle of $\mathbb{R}P^{n+1}$ defined as $K_-(x, v) = \min_{\sigma \subset T_x\mathbb{R}P^{n+1}} K(\sigma, v)$ and $K_+(x, v) = \max_{\sigma \subset T_x\mathbb{R}P^{n+1}} K(\sigma, v)$, where σ runs over all the planes in $T_x\mathbb{R}P^{n+1}$ and $K(\sigma, v)$ is the flag curvature. Notice that $K_- = K_+$ if $n = 1$.

Theorem 63 ([8]) *Let F be a Finsler metric on $\mathbb{R}P^{n+1}$ with $n \geq 1$, reversibility r and flag curvature K satisfying $(r/(r+1))^2 \leq K \leq 1$. If K is not strictly bigger than $(r/(r+1))^2$, suppose that $K_-(x, v)/K_+(x, v) > (r/(r+1))^2$ for every (x, v) in the unit sphere bundle of $\mathbb{R}P^{n+1}$. Then F carries an elliptic closed geodesic γ. Moreover, γ is prime and $\mu_{Morse}(\gamma) = 0$.*

Remark 64 The hypothesis that $K_-(x,v)/K_+(x,v) > (r/(r+1))^2$ allows us to make a C^∞-perturbation of F such that the flag curvature K' of the new metric F' satisfies $(r/(r+1))^2 < K' \le 1$; see [8, Proposition 6.1]. This, together with a uniform bound for the period of the closed orbit given by Theorem 50, enables us to deal with the case that the pinching condition is not strict. An analogous assumption appears in the next theorem by the same reason.

Under the stronger pinching condition $\frac{9}{4}(r/(r+1))^2 < K \le 1$, Rademacher [76] proved that every closed geodesic γ of F satisfies $\mu(\gamma^2) \ge 3n = 2\mu_{RS}(\varphi) - n$. Hence, by Theorem 58, we have the following generalization of [16, Theorem A(i)] proved before by Rademacher [77] using different methods and assuming that the pinching condition is strict. Similarly to the previous theorem, in the statement K_- and K_+ are defined as $K_-(x,v) = \min_{\sigma\subset T_xS^{n+1}} K(\sigma,v)$ and $K_+(x,v) = \max_{\sigma\subset T_xS^{n+1}} K(\sigma,v)$, where σ runs over all the planes in T_xS^{n+1}.

Theorem 65 ([8]) *Let F be a Finsler metric on S^{n+1} with $n \ge 1$, reversibility r and flag curvature K satisfying $\frac{9}{4}(r/(r+1))^2 \le K \le 1$. If K is not strictly bigger than $\frac{9}{4}(r/(r+1))^2$, suppose that $K_-(x,v)/K_+(x,v) > \frac{9}{4}(r/(r+1))^2$ for every (x,v) in the unit sphere bundle of S^{n+1}. Then F carries an elliptic closed geodesic γ such that $\mu_{Morse}(\gamma) \le n$. Moreover, if $\frac{9}{4}(r/(r+1))^2 < K \le 1$ then γ is prime and $\mu_{Morse}(\gamma) = n$.*

Our next result applies to magnetic flows and motivates our definition of *negative* dynamical convexity. Let N be a Riemannian manifold and Ω a closed 2-form on N. Consider on T^*N the twisted symplectic form $\omega = \omega_0 + \pi^*\Omega$, where ω_0 is the canonical symplectic form and $\pi : T^*N \to N$ is the projection. Let $H : T^*N \to \mathbb{R}$ be the Hamiltonian given by the kinetic energy. The Hamiltonian flow of H with respect to ω is called the magnetic flow generated by the Riemannian metric and the magnetic field Ω.

It is well known that if N is a closed orientable surface with genus $\mathfrak{g} \ne 1$ and Ω is a symplectic form then there exists $\epsilon > 0$ such that $H^{-1}(k)$ is of contact type for every $k < \epsilon$, see [17, 46] (an easy argument using the Gysin sequence shows that it cannot be of contact type if $N = T^2$). Benedetti proved in [17] that if $N = S^2$ then every periodic orbit γ of the lifted Reeb flow on S^3 satisfies $\mu(\gamma) \ge 3$ if ϵ is chosen sufficiently small.

Moreover, it follows from his estimates that, for surfaces of genus $\mathfrak{g} > 1$, a homologically trivial periodic orbit γ on $H^{-1}(k)$ satisfies $\mu^+(\gamma) \le 2\chi(N) + 1$, where $\chi(N)$ stands for the Euler characteristic of N and the index is computed using a suitable global trivialization of the contact structure; see [8, Sect. 6.5] for details. Now, fix $k < \epsilon$ and consider $S := H^{-1}(k)$ with the contact form α. One can prove that S is a prequantization and there is a $|\chi(N)|$-covering $\tau : \widetilde{S} \to S$ such that $\widetilde{\beta} := \tau^*\beta$ (where β is the contact form whose Reeb flow is periodic) generates a free circle action and the deck transformations are given by the induced action of $\mathbb{Z}_{|\chi(N)|} \subset S^1$. Moreover, the orbits of this circle action in $\widetilde{S}$ are homologous to zero. Let $\widetilde{\varphi}$ be a (simple) orbit of the Reeb flow of $\widetilde{\beta}$. An easy computation shows that $\mu_{RS}(\widetilde{\varphi}) = 2\chi(N)$.

Consequently, the lift of α to $\widetilde{S}$ furnishes a $\mathbb{Z}_{|\chi(N)|}$-invariant contact form satisfying the hypothesis (H2) in Theorem 50.

In other words, the magnetic flow on a sufficiently small energy level has a lift to a $|\chi(N)|$-covering that is *positively* dynamically convex if $\mathfrak{g} = 0$ and *a-negatively* dynamically convex if $\mathfrak{g} > 1$, where a is the free homotopy class of $\widetilde{\varphi}$. Therefore, we obtain the following result.

Theorem 66 ([8]) *Let (N, g) be a closed orientable Riemannian surface of genus $\mathfrak{g} \neq 1$ and Ω a symplectic magnetic field on N. There exists $\epsilon > 0$ such that the magnetic flow has an elliptic closed orbit γ on $H^{-1}(k)$ for every $k < \epsilon$. Moreover, γ is prime, freely homotopic to a fiber of SN and satisfies $\mu(\gamma) = 1$ if $\mathfrak{g} = 0$ and $\mu^+(\gamma) = -1$ if $\mathfrak{g} > 1$, where the indexes are computed using a suitable global trivialization of ξ over $H^{-1}(k)$.*

4 Symmetric Periodic Reeb Orbits

As mentioned in Sect. 2.2, a long standing conjecture in Hamiltonian Dynamics establishes that every contact form α on S^{2n+1} has at least $n+1$ simple closed Reeb orbits. In [6], it is considered a refinement of this conjecture when the contact form has a symmetry. More precisely, given an integer $p > 0$, consider the $\mathbb{Z}_p$-action on S^{2n+1}, regarded as a subset of $\mathbb{C}^{n+1} \setminus \{0\}$, generated by the map

$$\psi(z_0, \ldots, z_n) = \left(e^{\frac{2\pi\sqrt{-1}\ell_0}{p}} z_0, e^{\frac{2\pi\sqrt{-1}\ell_1}{p}} z_1, \ldots, e^{\frac{2\pi\sqrt{-1}\ell_n}{p}} z_n\right),$$

where $\ell_0, \ldots, \ell_n$ are integers called the *weights* of the action. Such an action is free when the weights are coprime with p and in that case we have a lens space obtained as the quotient of S^{2n+1} by the action of $\mathbb{Z}_p$. We denote such a lens space by $L_p^{2n+1}(\ell_0, \ell_1, \ldots, \ell_n)$. From now on, we will assume that the weights are coprime with p.

A natural question, addressed in [6], is the following one. Recall that given the action of a Lie group G on a manifold M and a vector field X on M invariant under this action, we say that a periodic orbit γ of X is *symmetric* if $g(\gamma(\mathbb{R})) = \gamma(\mathbb{R})$ for every $g \in G$.

Question
Let α be a contact form on S^{2n+1} invariant under the above $\mathbb{Z}_p$-action generated by ψ. Denote by $\mathcal{P}_s$ the set of simple *symmetric* closed orbits of the Reeb flow of α. Is it true that $\#\mathcal{P}_s \geq n+1$?

Note that when $p = 1$ the $\mathbb{Z}_p$-action is trivial and therefore every periodic orbit is symmetric. Thus, in this case the question is equivalent to the aforementioned problem if every contact form on S^{2n+1} carries at least $n+1$ closed orbits. Moreover, the irrational ellipsoids are invariant under this $\mathbb{Z}_p$-action for any p and weights $\ell_0, \dots, \ell_n$ and carry precisely $n+1$ periodic orbits which are all symmetric.

Girardi proved in [51] that when $p = 2$, that is, when ψ is the antipodal map, we have that $\#\mathcal{P}_s \geq 1$ for any symmetric contact form. In the same work, it was proved that if Σ_α satisfies suitable pinching conditions then $\#\mathcal{P}_s \geq n+1$ when $p = 2$. This result was generalized by Abreu and the author in [11, Theorem 1.21] for any p assuming that α is convex and $\ell_0 = \cdots = \ell_n = 1$. In [15], Bähni showed that $\#\mathcal{P}_s \geq 1$ for any symmetric contact form when p is even.

When α is convex and symmetric, for any p and weights $\ell_0, \dots, \ell_n$, it was proved by Zhang [89] that $\#\mathcal{P}_s \geq 2$. This result was generalized to dynamically convex contact forms by Liu and Zhang [62] assuming that $p = 2$.

The main results in [6] are the following generalizations of these results.

Theorem 67 ([6]) *Let α be any contact form on S^{2n+1} invariant under the above $\mathbb{Z}_p$-action. Then α has at least one symmetric closed orbit.*

This result generalizes the results of Rabinowitz [73] (that corresponds to the case $p = 1$), Girardi [51] (that corresponds to the case $p = 2$) and Bähni [15] (that corresponds to the case that p is even).

Theorem 68 ([6]) *Let α be a dynamically convex contact form on S^{2n+1} invariant under the above $\mathbb{Z}_p$-action. Then α has at least two simple symmetric closed orbits.*

The last result generalizes the results of Zhang [89] (that corresponds to the case that α is convex) and Liu-Zhang [62] (that corresponds to the case $p = 2$).

Let β be the induced contact form on $L_p^{2n+1}(\ell_0, \dots, \ell_n)$. A natural question is how the previous results can be seen in terms of periodic orbits of β. The simple symmetric closed orbits of α are in bijection with the simple closed orbits of β whose homotopy classes are generators of $\pi_1(L_p^{2n+1}(\ell_0, \dots, \ell_n))$. Thus, Theorem 67 is equivalent to the statement that given any $a \in \pi_1(L_p^{2n+1}(\ell_0, \dots, \ell_n))$ then every contact form β on $L_p^{2n+1}(\ell_0, \dots, \ell_n)$ has at least one periodic orbit with homotopy class a. Theorem 68, in turn, is equivalent to the statement that given any $a \in \pi_1(L_p^{2n+1}(\ell_0, \dots, \ell_n))$ and a dynamically convex contact form β on $L_p^{2n+1}(\ell_0, \dots, \ell_n)$ (we say that β is dynamically convex if so is its lift to S^{2n+1}; equivalently, β is 0-dynamically convex in the sense of Definition 10) we have at least two simple periodic orbits with homotopy class a, where a periodic orbit with homotopy class a is called simple if it is not an iterate of another periodic orbit with homotopy class a (although it can be the iterate of a closed orbit with another homotopy class).

References

1. Abbondandolo, A., Schwarz, M.: On the Floer homology of cotangent bundles. Comm. Pure Appl. Math. **59**(2), 254–316 (2006). https://doi.org/10.1002/cpa.20090
2. Abbondandolo, A., Schwarz, M.: Corrigendum: On the Floer homology of cotangent bundles. Comm. Pure Appl. Math. **67**(4), 670–691 (2014). https://doi.org/10.1002/cpa.21508
3. Abouzaid, M.: Symplectic cohomology and Viterbo's theorem. In: Latschev, J., Oancea, A. (eds.) Free Loop Spaces in Geometry and Topology, IRMA Lect. Math. Theor. Phys., vol. 24, pp. 271–485. Eur. Math. Soc., Zürich (2015)
4. Abraham, R., Marsden, J.E.: Foundations of mechanics, 2nd edn. Benjamin/Cummings Publishing Co., Inc., Advanced Book Program, Reading, MA (1978). With the assistance of Tudor Raţiu and Richard Cushman
5. Abreu, M., Gutt, J., Kang, J., Macarini, L.: Two closed orbits for non-degenerate Reeb flows. Math. Proc. Cambridge Philos. Soc. **170**(3), 625–660 (2021). https://doi.org/10.1017/S0305004120000018
6. Abreu, M., Liu, H., Macarini, L.: Symmetric periodic Reeb orbits on the sphere. Trans. Amer. Math. Soc. **377**(9), 6751–6770 (2024). https://doi.org/10.1090/tran/9216
7. Abreu, M., Macarini, L.: Contact homology of good toric contact manifolds. Compos. Math. **148**(1), 304–334 (2012). https://doi.org/10.1112/S0010437X11007044
8. Abreu, M., Macarini, L.: Dynamical convexity and elliptic periodic orbits for Reeb flows. Math. Ann. **369**(1-2), 331–386 (2017). https://doi.org/10.1007/s00208-017-1532-4
9. Abreu, M., Macarini, L.: Multiplicity of periodic orbits for dynamically convex contact forms. J. Fixed Point Theory Appl. **19**(1), 175–204 (2017). https://doi.org/10.1007/s11784-016-0348-2
10. Abreu, M., Macarini, L.: On the mean Euler characteristic of Gorenstein toric contact manifolds. Int. Math. Res. Not. IMRN **2020**(14), 4465–4495 (2020). https://doi.org/10.1093/imrn/rny151
11. Abreu, M., Macarini, L.: Dynamical implications of convexity beyond dynamical convexity. Calc. Var. Partial Differ. Equ.**61**(3), Paper No. 116, 47 pp. (2022). https://doi.org/10.1007/s00526-022-02228-1
12. Abreu, M., Macarini, L.: Periodic orbits of non-degenerate lacunary contact forms on prequantization bundles. `arXiv: /2406.08462` (2024). https://arxiv.org/abs/2406.08462
13. Arnaud, M.C.: Existence d'orbites périodiques complètement elliptiques des Hamiltoniens convexes présentant certaines symétries. C. R. Acad. Sci. Paris Sér. I Math. **328**(11), 1035–1038 (1999). https://doi.org/10.1016/S0764-4442(99)80320-6
14. Arnol′d, V.I.: Mathematical methods of classical mechanics, Graduate Texts in Mathematics, vol. 60, 2nd edn. Springer, New York (1989). https://doi.org/10.1007/978-1-4757-2063-1. Translated from the Russian by K. Vogtmann and A. Weinstein
15. Bähni, Y.: First steps in twisted Rabinowitz-Floer homology. J. Symplectic Geom. **21**(1), 111–158 (2023). https://doi.org/10.4310/jsg.2023.v21.n1.a3
16. Ballmann, W., Thorbergsson, G., Ziller, W.: Closed geodesics on positively curved manifolds. Ann. Math. (2) **116**(2), 213–247 (1982). https://doi.org/10.2307/2007062
17. Benedetti, G.: The contact property for symplectic magnetic fields on S^2. Ergodic Theory Dynam. Syst. **36**(3), 682–713 (2016). https://doi.org/10.1017/etds.2014.82
18. Besse, A.L.: Manifolds all of whose geodesics are closed, Ergebnisse der Mathematik und ihrer Grenzgebiete [Results in Mathematics and Related Areas], vol. 93. Springer, Berlin-New York (1978)
19. Bourgeois, F., Oancea, A.: Symplectic homology, autonomous Hamiltonians, and Morse-Bott moduli spaces. Duke Math. J. **146**(1), 71–174 (2009). https://doi.org/10.1215/00127094-2008-062
20. Bourgeois, F., Oancea, A.: Fredholm theory and transversality for the parametrized and for the S^1-invariant symplectic action. J. Eur. Math. Soc. (JEMS) **12**(5), 1181–1229 (2010). https://doi.org/10.4171/JEMS/227
21. Bourgeois, F., Oancea, A.: The Gysin exact sequence for S^1-equivariant symplectic homology. J. Topol. Anal. **5**(4), 361–407 (2013). https://doi.org/10.1142/S1793525313500210

22. Bourgeois, F., Oancea, A.: The index of Floer moduli problems for parametrized action functionals. Geom. Dedicata **165**, 5–24 (2013). https://doi.org/10.1007/s10711-012-9763-8
23. Bourgeois, F., Oancea, A.: S^1-equivariant symplectic homology and linearized contact homology. Int. Math. Res. Not. IMRN **2017**(13), 3849–3937 (2017). https://doi.org/10.1093/imrn/rnw029
24. Chaidez, J., Edtmair, O.: 3D convex contact forms and the Ruelle invariant. Invent. Math. **229**(1), 243–301 (2022). https://doi.org/10.1007/s00222-022-01107-y
25. Chaidez, J., Edtmair, O.: The ruelle invariant and convexity in higher dimensions. `arXiv: /2205.00935` (2022). https://arxiv.org/abs/2205.00935
26. Cieliebak, K., Latschev, J.: The role of string topology in symplectic field theory. In: Abreu, M., Lalonde, F., Polterovich, L. (eds.) New perspectives and challenges in symplectic field theory, CRM Proc. Lecture Notes, vol. 49, pp. 113–146. Amer. Math. Soc., Providence, RI (2009). https://doi.org/10.1090/crmp/049/04
27. Çineli, E., Ginzburg, V.L., Gürel, B.Z.: Closed orbits of dynamically convex reeb flows: towards the hz-and multiplicity conjectures. `arXiv: /2410.13093` (2024). https://arxiv.org/abs/2410.13093
28. Cristofaro-Gardiner, D., Hryniewicz, U., Hutchings, M., Liu, H.: Proof of hofer-wysocki-zehnder's two or infinity conjecture. `arXiv: /2310.07636` (2024). https://arxiv.org/abs/2310.07636
29. Cristofaro-Gardiner, D., Hutchings, M.: From one Reeb orbit to two. J. Differ. Geom. **102**(1), 25–36 (2016). http://projecteuclid.org/euclid.jdg/1452002876
30. Cristofaro-Gardiner, D., Hutchings, M., Pomerleano, D.: Torsion contact forms in three dimensions have two or infinitely many Reeb orbits. Geom. Topol. **23**(7), 3601–3645 (2019). https://doi.org/10.2140/gt.2019.23.3601
31. Croke, C.B., Weinstein, A.: Closed curves on convex hypersurfaces and periods of nonlinear oscillations. Invent. Math. **64**(2), 199–202 (1981). https://doi.org/10.1007/BF01389167
32. Dell'Antonio, G., D'Onofrio, B., Ekeland, I.: Periodic solutions of elliptic type for strongly nonlinear Hamiltonian systems. In: Hofer, H., Taubes, C.H., Weinstein, A., Zehnder, E. (eds.) The Floer memorial volume, Progr. Math., vol. 133, pp. 327–333. Birkhäuser, Basel (1995). https://doi.org/10.1007/978-3-0348-9217-9_14
33. Duan, H., Liu, H.: Multiplicity and ellipticity of closed characteristics on compact star-shaped hypersurfaces in $\mathbb{R}^{2n}$. Calc. Var. Partial Differ. Equ. **56**(3), Paper No. 65, 30 pp. (2017). https://doi.org/10.1007/s00526-017-1173-1
34. Duan, H., Liu, H., Long, Y.M., Wang, W.: Non-hyperbolic closed characteristics on non-degenerate star-shaped hypersurfaces in $\mathbb{R}^{2n}$. Acta Math. Sin. (Engl. Ser.) **34**(1), 1–18 (2018). https://doi.org/10.1007/s10114-016-6019-9
35. Duan, H., Liu, H., Ren, W.: A dichotomy result for closed characteristics on compact star-shaped hypersurfaces in $\mathbb{R}^{2n}$. Math. Z. **302**(2), 743–757 (2022). https://doi.org/10.1007/s00209-022-03085-6
36. Duan, H., Long, Y., Wang, W.: The enhanced common index jump theorem for symplectic paths and non-hyperbolic closed geodesics on Finsler manifolds. Calc. Var. Partial Differ. Equ. **55**(6), Art. 145, 28 pp. (2016). https://doi.org/10.1007/s00526-016-1075-7
37. Duan, H., Xie, D.: Multiplicity of closed geodesics on bumpy Finsler manifolds with elliptic closed geodesics. J. Funct. Anal. **284**(8), Paper No. 109861, 20 pp. (2023). https://doi.org/10.1016/j.jfa.2023.109861
38. Ekeland, I.: An index theory for periodic solutions of convex Hamiltonian systems. In: Browder, F.E. (ed.) Nonlinear functional analysis and its applications, Part 1 (Berkeley, Calif., 1983), Proc. Sympos. Pure Math., vol. 45, Part 1, pp. 395–423. Amer. Math. Soc., Providence, RI (1986). https://doi.org/10.1090/pspum/045.1/843575
39. Ekeland, I.: Convexity methods in Hamiltonian mechanics, Ergebnisse der Mathematik und ihrer Grenzgebiete (3) [Results in Mathematics and Related Areas (3)], vol. 19. Springer, Berlin (1990). https://doi.org/10.1007/978-3-642-74331-3
40. Ekeland, I., Hofer, H.: Convex Hamiltonian energy surfaces and their periodic trajectories. Comm. Math. Phys. **113**(3), 419–469 (1987). http://projecteuclid.org/euclid.cmp/1104160288

41. Eliashberg, Y., Givental, A., Hofer, H.: Introduction to Symplectic Field Theory. In: Alon, N., Bourgain, J., Connes, A., Gromov, M., Milman, V. (eds.) GAFA 2000. Visions in mathematics—Towards 2000. Proceedings of a meeting, Tel Aviv, Israel, August 25–September 3, 1999. Part II., pp. 560–673. Basel, Birkhäuser (2000). https://doi.org/10.1007/978-3-0346-0425-3_4
42. Ginzburg, V.L.: The Conley conjecture. Ann. Math. (2) **172**(2), 1127–1180 (2010). https://doi.org/10.4007/annals.2010.172.1129
43. Ginzburg, V.L., Gürel, B.s.Z.: On the generic existence of periodic orbits in Hamiltonian dynamics. J. Mod. Dyn. **3**(4), 595–610 (2009). https://doi.org/10.3934/jmd.2009.3.595
44. Ginzburg, V.L., Gürel, B.s.Z.: The Conley conjecture and beyond. Arnold Math. J. **1**(3), 299–337 (2015). https://doi.org/10.1007/s40598-015-0017-3
45. Ginzburg, V.L., Gürel, B.s.Z.: Lusternik-Schnirelmann theory and closed Reeb orbits. Math. Z. **295**(1-2), 515–582 (2020). https://doi.org/10.1007/s00209-019-02361-2
46. Ginzburg, V.L., Gürel, B.s.Z., Macarini, L.: On the Conley conjecture for Reeb flows. Internat. J. Math. **26**(7), 1550047, 22 (2015). https://doi.org/10.1142/S0129167X15500470
47. Ginzburg, V.L., Gürel, B.s.Z., Macarini, L.: Multiplicity of closed Reeb orbits on prequantization bundles. Israel J. Math. **228**(1), 407–453 (2018). https://doi.org/10.1007/s11856-018-1769-y
48. Ginzburg, V.L., Hein, D., Hryniewicz, U.L., Macarini, L.: Closed Reeb orbits on the sphere and symplectically degenerate maxima. Acta Math. Vietnam. **38**(1), 55–78 (2013). https://doi.org/10.1007/s40306-012-0002-z. Proceedings of the congress "Geometrical methods in Dynamics and Topology (celebrating the 60th anniversary of the Hanoi National University of Education)"
49. Ginzburg, V.L., Macarini, L.: Dynamical convexity and closed orbits on symmetric spheres. Duke Math. J. **170**(6), 1201–1250 (2021). https://doi.org/10.1215/00127094-2020-0097
50. Ginzburg, V.L., Shon, J.: On the filtered symplectic homology of prequantization bundles. Internat. J. Math. **29**(11), 1850071, 35 (2018). https://doi.org/10.1142/S0129167X18500714
51. Girardi, M.: Multiple orbits for Hamiltonian systems on starshaped surfaces with symmetries. Ann. Inst. H. Poincaré Anal. Non Linéaire **1**(4), 285–294 (1984). http://www.numdam.org/item?id=AIHPC_1984__1_4_285_0
52. Gromoll, D., Meyer, W.: Periodic geodesics on compact riemannian manifolds. J. Differ. Geom. **3**, 493–510 (1969). http://projecteuclid.org/euclid.jdg/1214429070
53. Gutt, J., Kang, J.: On the minimal number of periodic orbits on some hypersurfaces in $\mathbb{R}^{2n}$. Ann. Inst. Fourier (Grenoble) **66**(6), 2485–2505 (2016). https://doi.org/10.5802/aif.3069
54. Harris, A., Paternain, G.P.: Dynamically convex Finsler metrics and J-holomorphic embedding of asymptotic cylinders. Ann. Global Anal. Geom. **34**(2), 115–134 (2008). https://doi.org/10.1007/s10455-008-9111-2
55. Hingston, N.: Subharmonic solutions of Hamiltonian equations on tori. Ann. Math. (2) **170**(2), 529–560 (2009). https://doi.org/10.4007/annals.2009.170.529
56. Hofer, H., Wysocki, K., Zehnder, E.: The dynamics on three-dimensional strictly convex energy surfaces. Ann. Math. (2) **148**(1), 197–289 (1998). https://doi.org/10.2307/120994
57. Hofer, H., Wysocki, K., Zehnder, E.: Polyfold and Fredholm theory, Ergebnisse der Mathematik und ihrer Grenzgebiete. 3. Folge. A Series of Modern Surveys in Mathematics [Results in Mathematics and Related Areas. 3rd Series. A Series of Modern Surveys in Mathematics], vol. 72. Springer, Cham (2021). https://doi.org/10.1007/978-3-030-78007-4
58. Hryniewicz, U.L., Macarini, L.: Local contact homology and applications. J. Topol. Anal. **7**(2), 167–238 (2015). https://doi.org/10.1142/S1793525315500119
59. Lerman, E.: Contact toric manifolds. J. Symplectic Geom. **1**(4), 785–828 (2003). https://doi.org/10.4310/jsg.2001.v1.n4.a6
60. Liu, C.G., Long, Y., Zhu, C.: Multiplicity of closed characteristics on symmetric convex hypersurfaces in $\mathbb{R}^{2n}$. Math. Ann. **323**(2), 201–215 (2002). https://doi.org/10.1007/s002089100257
61. Liu, H., Long, Y.M.: The existence of two closed characteristics on every compact star-shaped hypersurface in $\mathbb{R}^4$. Acta Math. Sin. (Engl. Ser.) **32**(1), 40–53 (2016). https://doi.org/10.1007/s10114-014-4108-1

62. Liu, H., Zhang, L.: Multiplicity of closed Reeb orbits on dynamically convex $\mathbb{R}P^{2n-1}$ for $n \geq 2$. Discrete Contin. Dyn. Syst. **42**(4), 1801–1816 (2022). https://doi.org/10.3934/dcds.2021172
63. Long, Y.: Precise iteration formulae of the Maslov-type index theory and ellipticity of closed characteristics. Adv. Math. **154**(1), 76–131 (2000). https://doi.org/10.1006/aima.2000.1914
64. Long, Y.: Index theory for symplectic paths with applications. Progress in Mathematics, vol. 207. Birkhäuser Verlag, Basel (2002). https://doi.org/10.1007/978-3-0348-8175-3
65. Long, Y., Zhu, C.: Closed characteristics on compact convex hypersurfaces in $\mathbb{R}^{2n}$. Ann. Math. (2) **155**(2), 317–368 (2002). https://doi.org/10.2307/3062120
66. McLean, M.: Local Floer homology and infinitely many simple Reeb orbits. Algebr. Geom. Topol. **12**(4), 1901–1923 (2012). https://doi.org/10.2140/agt.2012.12.1901
67. McLean, M.: Reeb orbits and the minimal discrepancy of an isolated singularity. Invent. Math. **204**(2), 505–594 (2016). https://doi.org/10.1007/s00222-015-0620-x
68. Newhouse, S.E.: Quasi-elliptic periodic points in conservative dynamical systems. Amer. J. Math. **99**(5), 1061–1087 (1977). https://doi.org/10.2307/2374000
69. Pardon, J.: An algebraic approach to virtual fundamental cycles on moduli spaces of pseudo-holomorphic curves. Geom. Topol. **20**(2), 779–1034 (2016). https://doi.org/10.2140/gt.2016.20.779
70. Pardon, J.: Contact homology and virtual fundamental cycles. J. Amer. Math. Soc. **32**(3), 825–919 (2019). https://doi.org/10.1090/jams/924
71. Poincaré, H.: Sur les lignes géodésiques des surfaces convexes. Trans. Amer. Math. Soc. **6**(3), 237–274 (1905). https://doi.org/10.2307/1986219
72. Pugh, C.C., Robinson, C.: The C^1 closing lemma, including Hamiltonians. Ergodic Theory Dynam. Syst. **3**(2), 261–313 (1983). https://doi.org/10.1017/S0143385700001978
73. Rabinowitz, P.H.: Periodic solutions of Hamiltonian systems. Comm. Pure Appl. Math. **31**(2), 157–184 (1978). https://doi.org/10.1002/cpa.3160310203
74. Rabinowitz, P.H.: A variational method for finding periodic solutions of differential equations. In: M.G. Crandall (ed.) Nonlinear evolution equations: Proc. Sympos., Univ. Wisconsin, Madison, Wis., October 17-19, 1977, Publication of the Mathematics Research Center, University of Wisconsin, vol. 40, pp. 225–251. Academic Press, New York-London (1978)
75. Rabinowitz, P.H.: Periodic solutions of a Hamiltonian system on a prescribed energy surface. J. Differ. Equ. **33**(3), 336–352 (1979). https://doi.org/10.1016/0022-0396(79)90069-X
76. Rademacher, H.B.: A sphere theorem for non-reversible Finsler metrics. Math. Ann. **328**(3), 373–387 (2004). https://doi.org/10.1007/s00208-003-0485-y
77. Rademacher, H.B.: Existence of closed geodesics on positively curved Finsler manifolds. Ergodic Theory Dynam. Syst. **27**(3), 957–969 (2007). https://doi.org/10.1017/S0143385706001064
78. Robbin, J., Salamon, D.: The Maslov index for paths. Topology **32**(4), 827–844 (1993). https://doi.org/10.1016/0040-9383(93)90052-W
79. Salamon, D., Zehnder, E.: Morse theory for periodic solutions of Hamiltonian systems and the Maslov index. Comm. Pure Appl. Math. **45**(10), 1303–1360 (1992). https://doi.org/10.1002/cpa.3160451004
80. Seidel, P.: A biased view of symplectic cohomology. In: Jerison, D., Mazur, B., Mrowka, T., Schmid, W., Stanley, R.P., Yau, S.T. (eds.) Current Developments in Mathematics, 2006, pp. 211–253. Int. Press, Somerville, MA (2008)
81. Taubes, C.H.: The Seiberg-Witten equations and the Weinstein conjecture. Geom. Topol. **11**, 2117–2202 (2007). https://doi.org/10.2140/gt.2007.11.2117
82. Vigué-Poirrier, M., Sullivan, D.: The homology theory of the closed geodesic problem. J. Differ. Geom. **11**(4), 633–644 (1976). http://projecteuclid.org/euclid.jdg/1214433729
83. Viterbo, C.: Functors and computations in Floer homology with applications. I. Geom. Funct. Anal. **9**(5), 985–1033 (1999). https://doi.org/10.1007/s000390050106
84. Wang, W.: Stability of closed characteristics on symmetric compact convex hypersurfaces in $\mathbb{R}^{2n}$. J. Math. Pures Appl. (9) **99**(3), 297–308 (2013). https://doi.org/10.1016/j.matpur.2012.06.014

85. Wang, W.: Irrationally elliptic closed characteristics on compact convex hypersurfaces in $\mathbb{R}^6$. J. Funct. Anal. **267**(3), 799–841 (2014). https://doi.org/10.1016/j.jfa.2014.05.014
86. Wang, W.: Closed characteristics on compact convex hypersurfaces in $\mathbb{R}^8$. Adv. Math. **297**, 93–148 (2016). https://doi.org/10.1016/j.aim.2016.03.044
87. Wang, W.: Existence of closed characteristics on compact convex hypersurfaces in $\mathbb{R}^{2n}$. Calc. Var. Partial Differ. Equ. **55**(1), Art. 2, 25 (2016). https://doi.org/10.1007/s00526-015-0945-8
88. Weinstein, A.: On the hypotheses of Rabinowitz' periodic orbit theorems. J. Differ. Equ. **33**(3), 353–358 (1979). https://doi.org/10.1016/0022-0396(79)90070-6
89. Zhang, D.: P-cyclic symmetric closed characteristics on compact convex P-cyclic symmetric hypersurface in $\mathbb{R}^{2n}$. Discrete Contin. Dyn. Syst. **33**(2), 947–964 (2013). https://doi.org/10.3934/dcds.2013.33.947
90. Ziller, W.: Geometry of the Katok examples. Ergodic Theory Dynam. Syst. **3**(1), 135–157 (1983). https://doi.org/10.1017/S0143385700001851

GPSR Compliance

The European Union's (EU) General Product Safety Regulation (GPSR) is a set of rules that requires consumer products to be safe and our obligations to ensure this.

If you have any concerns about our products, you can contact us on ProductSafety@springernature.com

In case Publisher is established outside the EU, the EU authorized representative is:

Springer Nature Customer Service Center GmbH
Europaplatz 3
69115 Heidelberg, Germany

Batch number: 10755442

Printed by Printforce, the Netherlands